eries is intended to supply analysis and to stimulate debate. Opinions will
petween authors; we claim only that the books are based on searching
ical study of topics which are important, not least because they cut across
ntional academic boundaries. They should appeal not just to historians,
st to scientists, engineers and doctors, but to all who share the view that
e, technology and medicine are far too important to be left out of history.

include:

ta E. Bivins
UNCTURE, EXPERTISE AND CROSS-CULTURAL MEDICINE

Cooter
ERY AND SOCIETY IN PEACE AND WAR
paedics and the Organization of Modern Medicine, 1880–1948

Edgerton
AND AND THE AEROPLANE
say on a Militant and Technological Nation

Paul Gaudillière and Ilana Löwy (*editors*)
INVISIBLE INDUSTRIALIST
facture and the Construction of Scientific Knowledge

ie Smith and Jon Agar (*editors*)
NG SPACE FOR SCIENCE
orial Themes in the Shaping of Knowledge

ce, Technology and Medicine in Modern History
Standing Order ISBN 0–333–71492–X hardcover
Standing Order ISBN 0–333–80340–X paperback
de North America only)

an receive future titles in this series as they are published by placing a standing order.
contact your bookseller or, in case of difficulty, write to us at the address below with
name and address, the title of the series and one of the ISBNs quoted above.

mer Services Department, Macmillan Distribution Ltd, Houndmills, Basingstoke,
shire RG21 6XS, England

Science, Technology and Medicine in Modern History

General Editor: **John V. Pickstone**, Centre for the Histo
and Medicine, University of Manchester, England (ww

One purpose of historical writing is to illuminate
twentieth century, science, technology and medicine ai
yet their development is little studied. Histories of polit
and historical biography is established as an effective v
in context. But the historical literature on science, tec
relatively small, and the better studies are rarely accessi
Too often one finds mere chronicles of progress, or scie
do little to illuminate either the science or the society i
let alone their interactions.

The reasons for this failure are as obvious as they are
many countries, not least in Britain, draws deep divisic
and the humanities. Men and women who have been tra
often been trained away from history, or from any sust
societies work. Those educated in historical or social stud
so little of science that they remain thereafter suspiciou

Such a diagnosis is by no means novel, nor is it particu
that good historical studies of science may be pe
understanding our present. Indeed this series could be se
undertaken over the last half-century, especially by A
much of that work has treated science, technology and r
series aims to draw them together, partly because th
become ever more intertwined. This breadth of focus
relationships of knowledge and practice are particularly
which will concentrate on modern history and o
Furthermore, while much of the existing historical scho
topics, this series aims to be international, encouragin
material. The intention is to present science, technology i
of modern culture, analysing their economic, social and p
neglecting the expert content which tends to distance th
of history. The books will investigate the uses and con
knowledge, and how it was shaped within particular
political structures.

Such analyses should contribute to discussions of pres
assessments of policy. 'Science' no longer appears to us as
Enlightenment, breaking the shackles of tradition, ena
nature. But neither is it to be seen as merely oppre
Judgement requires information and careful analysis, jus
making requires a community of discourse between men
technical specialities and those who are not.

Acupuncture, Expertise and Cross-Cultural Medicine

Roberta E. Bivins
Wellcome Research Associate
Centre for the History of Science, Technology and Medicine
University of Manchester

palgrave

in association with
Centre for the History of Science,
Technology and Medicine,
University of Manchester

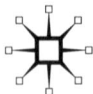

First published 2000 by
PALGRAVE
Houndmills, Basingstoke, Hampshire RG21 6XS and
175 Fifth Avenue, New York, N. Y. 10010
Companies and representatives throughout the world

PALGRAVE is the new global academic imprint of
St. Martin's Press LLC Scholarly and Reference Division and
Palgrave Publishers Ltd (formerly Macmillan Press Ltd).

ISBN 0–333–91893–2

This book is printed on paper suitable for recycling and made from fully managed and sustained forest sources.

A catalogue record for this book is available from the British Library.

Library of Congress Cataloging-in-Publication Data
Bivins, Roberta E., 1970–
    Acupuncture, expertise, and cross-cultural medicine / Roberta E. Bivins.
        p. cm. — (Science, technology, and medicine in modern history)
    Includes bibliographical references and index.
    ISBN 0–333–91893–2
    1. Acupuncture—Great Britain—History. 2. Transcultural medical care—
    —Great Britain. I. Title. II. Series.
    RM184 .B55 2000
    615.8'92'0941—dc21
                                                            00–033326

10   9   8   7   6   5   4   3   2   1
09 08 07 06 05 04 03 02 01 00

Printed and bound in Great Britain by
Antony Rowe Ltd, Chippenham, Wiltshire

To my colleague, mentor, and author: Mary Wren Bivins

And with what interest will a future generation peruse these concentrated records of the medical science of the present aera, distinguished as it is, by intensity of thought, accuracy of observation, rigour of deduction, intrepidity of experiment, fertility of resource, and emancipation from prejudice!

The Medico-Chirurgical Review, 1820

Contents

List of Illustrations

All the illustrations are reproduced courtesy of the Wellcome Trust Medical Photographic Library.

Acknowledgements

I began my research on the history of acupuncture in Britain under the pleasant delusion that it began in 1971, and would provide a nice introduction to a detailed analysis of the contemporary use of therapeutic needling and traditional Chinese medicine in the UK. Since coming to my senses with a bang some time in 1993, I have accumulated countless debts of gratitude to the scholars known and unknown who have made this history possible. First and foremost, I must thank the patient souls who guided me through my dissertation: Harriet Ritvo, the supervisor every student hopes for, and the generous and creative reader every scholar needs; Allan Brandt, without whose encouragement I would not have begun (much less finished) this project; and Evelynn Hammonds, whose rigorous and insightful reading helped me discover the meaning underneath the story. Roy Porter introduced me to the riches of the Wellcome Library and the Wellcome Institute for the History of Medicine, and has subsequently advised me on everything from scholarship to publication; the bibliography of this volume only begins to illustrate my scholarly debt to him. Rima Apple, Londa Schiebinger and Andrew Wear cheered me on (and cheered me up), sharing juicy sources in the parched hours of 'preparing for publication'. Kan-Wen Ma's deep knowledge of acupuncture's history and practice – not to mention his unfailing kindness and collegial generosity – saved me from many errors and steeled me to face the challenge of Chinese medicine. My time at the Wellcome Institute was enriched by conversations with many others; to name a few: David Allan, Sue Ferry, Kim Pelis, Cheryce Kramer, Bridie Andrews, Rhodri Hayward, Alan Yoshioka, Gita Tailor and Sally Bragg, and the fantastic librarians and archivists of the Wellcome Library. Dominic Wujastyk, in particular, answered countless questions about South Asian medicine with aplomb and erudition.

I owe a very special debt to staff at Lewisham Hospital's Health Services Research and Evaluation Unit and Complementary Therapy and Research Centre, and especially to Janet Richardson. Without her early enthusiasm, her dedicated and rigorous research, and her help in collecting information about the Centre and about complementary medicine in the National Health Service, my task in writing about twentieth-century British acupuncture would have been much more difficult (and far less fun). I am also grateful to Richard Hughes-Rowlands, of the

Special Trustees of the General Infirmary at Leeds for kindly permitting me to use the Infirmary's archives, and for his hospitality to me while I worked there.

Since coming to Manchester, I have benefited enormously from John Pickstone's advice and encouragement, his deep knowledge of medicine in Britain, and many (many!) invaluable arguments about the nature and process of making history. My colleagues, Claudia Castañeda, Lyn Schumaker and Helen Valier made me feel at home – and forced me talk about theory, for which I am immeasurably grateful. This book would have been much the poorer without Ian Burney's generosity with last-minute readings and insights. Thanks also to Penny Gouk, Roger Cooter, Steve Sturdy and others here in the beautiful North.

Without the financial support of the Wellcome Trust, both to me as an individual and to the discipline of the history of medicine, this book would never have been written. Thanks are due also to the Wellcome Trust Medical Photographic Library for kindly permitting me to reproduce crucial images from rare seventeenth- and eighteenth-century texts. Similarly, I must thank the *History of Science* for allowing me to reprint a slightly revised version of 'Expectations and Expertise: Early British Responses to Chinese Medicine and Technology', *History of Science* 37 (1999) here as Chapter 1; and the *Journal for the History of Medicine and Allied Sciences* for allowing me to quote (by permission of Oxford University Press) at length from Robert Carrubba and John Z. Bowers, 'The Western World's First Detailed Treatise on Acupuncture: Willem Ten Rhijne's *De Acupunctura*', *Journal of the History of Medicine*, 29 (1974): 371–98.

Finally I take great pleasure in thanking the family and friends who put up with me through dissertations, articles and the endless frustrations of re-writing. Such thanks can never repay but can only acknowledge them for their support, intelligence, patience and love: Mary, to whom this book is dedicated; Linda, my sister, who managed not to kill me; my grandparents; Julie; Carolyn; Dave; Kate; Rebecca; Betty; John and Toni; Jenny; the Mabey family; Maria; Aya; and perhaps most deservedly of all, the brilliant and charming Dr Mouse, who has kept me sane, fed me chip butties, and read me the book of the twentieth century.

Introduction: Cultural Specificity and the Cross-Cultural Transmission of Expert Knowledge

> It is highly probable that a few practical men admitted among them would in a few weeks acquire a mass of information for which if placed in the industrious and active hands of English manufacturers the whole revenue of the Chinese Empire would not be thought sufficient equivalent.[1]
>
> Joseph Banks, 1793

Medicine, as a set of practices and as an intellectual discipline, reflects in intimate detail the culture in which it develops; medicine as a profession necessarily partakes in its host culture's dominant values. These two insights have been at the heart of recent studies in the history of medicine, and have exposed many previously disregarded factors underlying medical continuity and change. And yet, medical practices and medical knowledge evidently do cross cultural boundaries, despite being rooted in specific understandings of the body determined by their native context. The focus on social construction which has complicated traditional medical narratives about cross-cultural transmission has also directed the historical gaze away from the issues central to that process. Ethnographic and sociological studies of the diffusion of medical innovation, and historical examinations of medicine's role in the expansion of empire are beginning to address the problem of cross-cultural transmission of medical knowledge. However, this body of work tends to focus on the spread of western ideas, technologies and practices to the non-western world.[2] Given the massive influx of consumer goods, ideas and technologies into Europe from Asia and the Middle East in the modern period, a unidirectional flow of medical expertise should be sharply contrary to scholarly expectations, rather than (as the current paucity of scholarship on the subject suggests) implicitly assumed.[3] The chapters which follow document the export of acupuncture from China and Japan to Great Britain, in order to explore the relationship between the cultural specificity of medicine and responses to medical innovation and exogenous medical expertise.[4]

The thread of exoticism – and its more scholarly analogue, orientalism – runs through the history of acupuncture in the West. That medicine, like other western cultural productions, was influenced by orientalist sensibilities is unsurprising (if somewhat under-examined). In this regard, western responses to acupuncture parallel responses to East Asian expertise and expert knowledge of all sorts from the mid-seventeenth century – immediately before the introduction of acupuncture to Europe – to the end of the Victorian era. But to what extent did the exotic origins of Asian innovations in technology and science – for example porcelain manufacture and tea-planting – affect their integration into British culture? And did responses to cross-cultural medical expertise follow the same pattern?

By the seventeenth century, Europeans were drawing distinctions between Chinese expertise in the 'mechanical arts' and Chinese abstract knowledge. While the latter was denigrated as useless, or used as evidence of China's inferiority to Europe, the former was advanced as an argument for further exploration and observation of Chinese culture. The returned missionary Louis LeComte, writing to the French Secretary of State, Philipeaux, included copious details of Chinese products and tools – emphasizing the skill and experience of Chinese artisans – and complained bitterly of Chinese reluctance to give up their manufacturing secrets. Tellingly, he contrasted their aptitude in the production of goods with its opposite in the production of knowledge:

> The Chinese that are mean proficients in the sciences, succeed much better in arts; ... they [know] in this respect not only what is necessary for the common use of life, but also whatever may contribute to the convenience, neatness, commerce, and even to well-regulated magnificence ... The workmen are extraordinary industrious, and if they be not so good at inventions as we, yet they do easily comprehend our inventions, and imitate them tolerably well ...[5]

However, distrust of Chinese learned knowledge, and of those who praised it only deepened over the course of the eighteenth century. By the time Britain's first official embassy to China set off in 1794, the demand for new, authentic – and British – accounts of the great empire was enormous. As we will see, the results of that embassy were unfavourable to Britain's trade, and China's reputation as a centre of civilization. John Barrow, who accompanied the mission, damned Chinese scholarship and natural knowledge conclusively with the comment that 'Having ascertained the fact, they have given themselves no further trouble to explain the phenomenon.'[6] Barrow commented extensively on China's crafts and manufactures; his remarks on Chinese porcelain are typical: 'The manufacture of earthenware, as far as depends upon the preparation of the materials, they have carried to a pitch of perfection not hitherto equalled by any nation except the Japanese ... [T]he beauty of their porcelains, in a great degree depends upon the extreme labour and attention that is paid to the assortment and the preparation of the different articles employed'.[7] The craft skills and diligence of the Chinese workman were indisputable, but they were not, Barrow argued, usefully transferable to British industry. Thus even when China's products competed successfully with their British counterparts, their excellence was based in effort rather than expert knowledge. Nor could such virtues fully compensate for China's deficiencies in rational thought: 'Neither the Chinese nor the Japanese can boast of giving to the materials much elegance of form. With the inimitable models from the Greek and Roman urns, brought into modern use by the ingenious Mr. Wedgewood, they will not bear a comparison.'[8] It is worth noting that British manufacturers, like Chinese artisans, were merely 'ingenious'. For Barrow, as would increasingly be the case in Britain, only producers of abstract knowledge could claim true 'genius'.

Other members of the mission were more optimistic; its official historian described China's many vast compendia of natural knowledge and concluded that: 'The Chinese books are full of the particular processes and methods, by which a variety of effects are produced in chemical and mechanical arts; and much might probably be gained from the perusal of them, by persons versed at the same time, in the language of the describers, and acquainted with the subject of the description.'[9] Certainly, this response was more positive than Barrow's. But such intellectual prospecting was in essence just an expansion of mercantilism. Chinese scholarly knowledge was to be mined for facts and marketable commodities in exactly the same way that China's soil might have been mined for minerals – and the facts thus collected were to be employed with as little attention to context and meaning as any metal ore.

From the turn of the nineteenth century on, knowledge about Chinese technology, manufactures and produce was indeed actively sought by Britons in the service of the Honourable East India Company, the British government, private and public institutions, and on their own initiative. Probably the most striking example of such activity is the extraction and transfer of the tea plant. Taking advantage of the opening of China's interior in the wake of the Opium Wars, the Honourable East India Company commissioned Robert Fortune to collect not only the tea plants, but also information about their cultivation and processing in China. Fortune was by no means the first to bring tea plants out of China alive; however, only he was able to observe Chinese agricultural techniques and tea processing. His detailed notes on every aspect of tea cultivation were rewarded by the plant's successful naturalization in India.[10] The potential commercial value of such knowledge was clearly expressed, and participants in official knowledge gathering operations, were, as David Mackay has demonstrated, well aware of the sensitivity of their missions.[11] Smuggling tea plants to India, surreptitiously collecting information about tea-processing, and creeping around the potteries to guess at the temperature of their kilns during glaze-firing were, after all, acts of industrial espionage. European accounts of China, too, were rapidly translated, lest competitive advantage be lost. Among them was a manuscript originally commissioned by yet another French minister of state:

[T]he missionaries, who were...so competent to elaborate history, philology, and mathematics...did not possess the same advantages in divining the secrets of arts and manufactures. In fact, the collections which the missionaries made were often defective and incomplete in many points of importance, owing either to their own ignorance, or to the disinclination of those to whom they were forced to apply. Besides, the Chinese manufacturers, equally jealous with our own, of the secrets of their respective professions, would not lightly disclose them to foreigners.[12]

Bertin commissioned his own Chinese spies at considerable expense; they were to focus on useful knowledge alone.

Such active and (as in the case of tea) organized pursuit of Chinese industrial expertise is radically different from British responses to Chinese medicine. Composed as it was of both theoretical and empirical expert knowledge, Chinese medicine provoked an ambivalent response. While Chinese anatomy was scorned, China's materia medica was lauded as a potential treasure trove: 'they draw the main part of their medical assistance from the long experienced virtues of the vegetable kind, from gentle purgatives, emollients, alteratives, and other salubrious remedies, calculated to strengthen, rather than fatigue and weaken . . . And it must be owned, that they have . . . the greatest plenty and variety of medicinal plants and roots, exactly suited for that purpose . . .'[13] It is revealing that Chinese medicine received only passing mentions in Bertin's four volumes – and perhaps even more so that only China's materia medica was considered worthy of further study. 'The Chinese . . . have an immense herbal, in two hundred and sixty volumes, from which our literary men would doubtless reap an ample harvest of new observations and discoveries.'[14] Although knowledge about local medical traditions was also sought by curious individuals (often medical missionaries who wanted to know their enemy, and to describe the advantages of western medicine in terms familiar to their native audiences) Chinese medical expertise was not considered worthy of pursuit in any organized way.

In the second half of the nineteenth century, with the forced opening of China to western residents, a new type of observer began to write about China. The orientalist had moved still further east, and scholarly attention began to focus on China's traditions. In medicine and technology, as well as the areas of philosophy, literature and religion, orientalist interest flourished. Unlike the cultural mercantilists who preceded them, the orientalists were explicitly not seeking gain – at least not material gain. Many saw themselves as continuing the mission which Barrow had set himself fifty years before – to establish the rank of China among civilized nations – with the additional goal of determining the cause for China's cultural decline.

Sources for the study of 'medical orientalism'

No scholar of the relationships between European and non-European expert cultures can avoid the implications of orientalism or the critical stance initiated (at least for my academic cohort) by Edward Said.[15] What I have called 'medical orientalism' has taken various forms. In the writings of the medical professional, China offered an opportunity to comment and exemplify upon medical debates at home. In much the same way that the Philosophes had used an idealized and glittering China as a lens through which to examine and condemn France, so well-travelled British medical men used Chinese medicine as a mirror to reflect – or perhaps to point out to an insufficiently appreciative audience – the glories of European medicine and culture. Thus medical accounts of China can reinforce or refute hypotheses about the history of medicine in Britain, as well as illuminating the perceived cultural impact of 'medicine' over time.

However, although we know that popular perceptions and expectations of medicine did change over the course of the nineteenth century, it has been more difficult to isolate the factors effecting that change. Was public opinion swayed by professional claims about education? By assertion about the value of the collateral sciences? Were medical consumers persuaded to value the profession by the scientificity or indeed efficacy of contemporary medical practice? Lay accounts of Chinese medicine and medical practitioners can shed light on these questions. What makes such popular accounts of China particularly interesting and useful is that they are written by laymen to appeal to a lay audience – and they did so very successfully. Books like *The Foreigner in Far Cathay* and *The Chinese as They Are* went through multiple printings, were comparatively inexpensive, and were used as authoritative sources of facts about China by newspapers, magazines and commentators in all but the most scholarly milieux.[16] Their authors were not medics, or indeed particularly engaged by the medical debates and reforms ongoing in Britain, though they were clearly influenced by those changes. Depictions of Chinese medicine were included in their narratives not because such descriptions offered a means by which to contribute to professional disputes at home, but because they were seen as illustrative of the Chinese character and civilization, and as informative about the Chinese mind, body and capabilities. The praise or condemnation these authors heap on Chinese medical and surgical practices, then, reveals as much about their expectations of western medicine and the social roles of medical practitioners as it does about the nature of Chinese healing. More importantly perhaps, these texts exploit comparisons of different medical systems and alternate views of the body in order to explicate some fundamental social and cultural debates about the nature and direction of British society in the nineteenth century. That medicine and the conduct of medical practitioners could play such a central role in debates about the very nature of civilization is in itself suggestive of their changing places in British culture. For lay authors, medicine proved an effective lens though which to examine culture more generally.

Despite the different perspectives of their authors, almost every lay account of Chinese medicine from this period shares two features: first, they describe Chinese medical knowledge and skill as in decay, degraded from a more glorious former state. Second, they compare it to European medicine of the more or less distant past. Though these themes do emerge in professional reports, they are by no means either so prominent or so nearly universal. Thus, J. F. Davis, the retiring Chief Superintendent of British trade in China, wrote in 1837:

> The actual state of the sciences in China may perhaps be ranked with their condition in Europe, some time previous to the adoption of the inductive method in philosophy...The jargon employed in their pseudo-science, and the singular resemblance which this bear to the condition of physical knowledge, not very long ago, even in our own country, is worthy of remark...They also have some vague notions of the *humoral pathology*, long since exploded in this country.[17]

The conclusion Davis drew from his observation is highly revealing:

> All this looks very much as if the philosophy of our forefathers was derived intermediately from China; and it is this easy plan, of *systematizing without experiment*, that has kept the latter country in the dark, and infested every department of its physical knowledge while the inductive philosophy recommended in the *novum organon* of Bacon has done such wonders in Europe.[18]

For authors like Davis, descriptions of medicine were doing another kind of work – they were illustrating the validity of recapitulation as the process by which civilizations (as well as species) evolved, and thus could be comparatively assessed. It is this use of Chinese medicine that best exemplifies 'medical orientalism'. Samuel Kidd, University College London's first Professor of Chinese language and literature, included a translation from a Chinese medical text among his early translations. He chose to translate the text literally, and in the language of metaphysics rather than that of his medical contemporaries. He followed the translated passage with the remark:

> I give this quotation, not with the hope of illuminating the faculty in Europe on the subject of medical treatment, but to show the superstitious notions and childish practices adopted by the most eminent Chinese physicians, who, during a long succession of ages, have had every opportunity of improving physical sciences which an extensive and populous country yielding the natural productions of every variety of climate and soil, was adapted to supply.[19]

The gaps in Chinese medical knowledge were evidence that China had failed to evolve as a culture, despite every material advantage. Such lay texts implicitly equate China's 'failure' to cultivate abstract knowledge with the failures of other indigenous peoples to appropriately exploit their natural resources. Either was sufficient to justify British economic and military intervention.

Less scholarly lay accounts were far more sympathetic to the notion of medicine as an art and a craft, rather than exclusively a science. The collected experience of centuries of Chinese medical men was frequently cited as a rationale for their undeniable therapeutic success, even though Chinese doctors were routinely chastized for their unshakable adherence to traditional knowledge and practice. Laymen were also more willing to act as cultural interpreters, translating Chinese systems and practices into terms explicable and even attractive to western audiences, rather than deliberately selecting for the exotic, metaphysical and incommunicable aspects. More importantly, they illustrate the place which medicine had attained as a measure of culture both in individuals and in civilizations.

By the mid-nineteenth century, two distinct patterns of response to Chinese medical knowledge had emerged in Britain, each of which found echoes in British reactions to other forms of Chinese expertise. Broadly speaking, the style of response I have called medical mercantilism emerged first, and waned first as well. After an interval of hostility during the period of the Opium Wars, it was gradually replaced by or developed into a style of response I have described here as medical orientalism, in which the intellectual and cultural products of China were no longer seen even as raw materials, but were considered worthy of scholarly investigation as revealing examples of the differences between orient and occident, past and present. Abstract knowledge, like understandings of the body, proved not to be portable. Western observers, whether professional or amateur, simply could not recognize in Chinese scholarship and scholarly learning, the apparatus and conclusions which they had come to associate with the pursuit of truth – and in the case of natural knowledge, with science. Similarly, although observers could see in the theoretical basis of Chinese medicine a likeness to humoral medicine, they could find no recognizable logic, no acceptable system by which to assimilate and evaluate it. Chinese learning, according to several commentators, was all madness and no method. As Frederic Balfour, scholar-translator of Taoist texts and author of a colloquial Chinese dictionary, complained in the 1880s:

There seem a looseness of reasoning, a want of consecutiveness in the mental processes of the Chinese which argues an inherent defect in their constitutions . . . The same phenomena is, of course, observed in all their so-called scientific theories. Physiology and metaphysics appear to form but one science according to Chinese notions, no clear distinction being recognised between phases of matter and phases of mind. This is almost incomprehensible to the European intellect; but it is none the less a fact.[20]

Yet Balfour continued to read and translate Chinese medical texts, noting that 'the Chinese are sufficiently well acquainted with the functions of the heart, and the relation to that organ of the blood. They know that for perfect health of body it is necessary that the blood should be kept completely pure.' Despite his lightly mocking tone, Balfour achieved an important insight into the differences between Chinese and western conceptions of the body:

In discussing metaphysical subjects with a cultured Chinese it is almost impossible to make him distinguish clearly between the physical organ [of the heart] and the word in its popular acceptation of mind . . . [W]e find the actual blood-pump made the seat or embodiment of the man's mental and moral characteristics; so much so, that every form of what we understand by the term heart-disease should logically be regarded as the sign of some special depravity or sin. Hence comes the curious Chinese doctrine of the effect of climate upon character . . .[21]

The history of acupuncture in the West has been shaped by the casual imposition of this European distinction between physic and metaphysics.

Overview

Although earlier reports of a therapy involving the insertion of needles into the body had reached the West, the first substantial discussion of acupuncture to arrive in Europe was written by Wilhelm Ten Rhyne, a physician employed by the Dutch East India Company, and stationed at their trading post in Japan. His essay *De Acupunctura*, written in Latin, was published in London in 1683. The second important work in the transmission process also emerged from Japan, and was written by one of Ten Rhyne's successors, Engelbert Kaempfer. Although both of these texts included maps of the traditional acupuncture points, and some brief discussion of East Asian medical theory, the European response to acupuncture focused on the needle and the techniques surrounding it. This emphasis on the material and mechanical aspects of acupuncture led to severe misinterpretations of the goals and practice of needling, and certainly hampered attempts to incorporate acupuncture into western medicine. However, by the end of the eighteenth century, a new group of practitioners was emerging in France. These men were experimentalists and clinicians; they took up acupuncture in part because it was a decidedly marginal practice, and thus available for experimental investigation in a way that established techniques were not.[22]

In the hands of the French clinicians, acupuncture proved successful in the treatment of ailments which orthodox western medicine had found intransigent. Reports both of the practical efficacy of acupuncture and of the experimental investigations of the needles were rapidly transmitted to Great Britain, and in 1822, the first significant English-language text on acupuncture was published by a surgeon named James Churchill. For a decade, the visibility and popularity of acupuncture increased steadily in Britain. By the mid-1830s, however, the number of articles on acupuncture had begun to decline. From 1850 until 1880, the medical press was virtually silent on the use of needles to alleviate pain, although other medical uses for the needle – some also called acupuncture – were discussed. Late in the century, analgesic acupuncture, this time particularly for neuralgia, re-surfaced in medical periodicals and tracts. However, as the twentieth century began, narratives of acupuncture disappeared from the medical literature; the technique of acupuncture was not again discussed in a medical context until the 1960s and 1970s.

The Anglo-European response to acupuncture was influenced by the politics and ideologies of the Enlightenment; by perceptions of China and Japan; by changes in the structure and membership of the medical professions and consequent shifts in the nature of medical practice; and by the growing authority of science and of scientific medicine. Because the history of acupuncture in Britain is composed of so many divergent strands, and because it is so little-known, I have organized the book as a loosely chronological narrative. But before turning to the first episode of transmission in 1683, I have made one detour. In Chapter 1, I have used the first formal British embassy to China in 1792 to explore the reasons for studying the cross-cultural transmission of medical knowledge. Lord Macartney, his colleagues, and their servants encountered Chinese medicine on several occasions and under a range of circumstances. Their interpretations of, and responses to these episodes are revealing of their expectations and opinions of medicine, as well as of China.

In Chapter 2, I do begin to address acupuncture, and the way in which information about the technique was accumulated and transmitted to Europe. The texts produced by Ten Rhyne and Kaempfer, and the images which accompanied them, were central to this process. Both men would have been members of the medical elite had they remained in Europe, and it was to the elite that their work was addressed. After examining these two first-hand accounts of acupuncture, the chapter continues by looking at lay and professional responses to the massive influx of information about Asian medicine in general and acupuncture in particular. This chapter also introduces a closely allied Asian medical practice, moxibustion, which was transmitted simultaneously. By comparing the reactions to these two exotic imports, the relative importance of their origins and of European medical expectations becomes more clearly apparent. The chapter ends with the century, and with the transmission of acupuncture, via the hospitals and experimental medicine of Paris, to Britain.

In Chapter 3, I have focused on the texts of the British response, and the 'acupuncture' which those texts constructed. Between 1822 and 1832, acupuncture received a level of exposure and examination which was not equalled in the West until the 1970s. Yet no theoretical explanation for acupuncture gained consensus. I have looked at the strategies used by proponents of acupuncture to cope with its resistance to categorization, and to popularize the therapeutic needle. They achieved a certain level of success; acupuncture was used in British hospitals as well as in private practice, and gained a place in the durable resources of British medicine. However, by the end of the decade its popularity and visibility were declining.

In Chapter 4, I examine the means by which acupuncture-use survived and even spread as the medical periodicals which had initially diffused the technique reduced, then ended their coverage of the topic. Predictably, acupuncture had become entangled in the internecine warfare which occupied the medical profession in these years of radicalism and reformist movements, and I consider which sections of the profession adopted the needle. I also explore some of the possible explanations for the submergence of acupuncture, both in the mid-century and – after its brief but intriguing fin-de-siècle resurgence – as the twentieth century began.

Inevitably, this remains a preliminary study of a complex and often puzzling episode in British and European medicine. The evidence presented here for acupuncture's transmission and British reception is drawn almost entirely from the printed record; few of the main actors in this story were survived by their correspondence or by the records of their businesses or medical practices. Of the few, neither patients nor practitioners recorded private opinions of, or encounters with, the healing needle.[23] Fortunately, the medical voice of the eighteenth and nineteenth centuries was more expressive than its modern counterpart. Medical authors were far more explicitly responsive to the social, intellectual and professional conflicts of the day, while the medical periodicals were politically and socially active. The rapid expansion of the medical periodicals in the nineteenth century also made the printed page accessible, both as a forum and as a professional resource, to a far wider range of medical practitioners. Moreover, the conventions of medical writing through the first half of the nineteenth century encouraged the inclusion of a patient's voice, although this construction was almost inevitably a filtered, formulaic – and easily fictionalized – shadow of the words of individual patients. All too often, of course, the intimate record of acupuncture in Britain is composed of gaping (but suggestively shaped) holes, bridged by only the slenderest of strands of evidence. However, through the formalized informalities – and the sheer volume – of medical communication, traces of the mundane, unpublished context of British acupuncture become visible.

Understanding the needle: tool, technique and techne

Academics conventionally discuss the complex unit formed by the culturally mediated interaction of individuals and implements in terms of one of two models. They may conceive of the tool and the techniques employing it as super-saturated with, and therefore inextricably embedded in, culture. Alternatively, the tool itself may be seen as culturally neutral (and thus completely transferable), with the techniques being considered as either embodied in the tool, or superfluous to it. In the first case, the expected outcome of cross-cultural transmission would be failure, unless the recipient culture adopted not only the tool, but all of the cultural constructs which surrounded it. The success of western surgical practices in China in the late nineteenth century, following the export of western medical schools, hospitals and practitioners, would be an example of this kind of successful context-rich transmission. In the second case, where the tool is conceived of as distinct from the culture within which it was developed, transmission would be of the decontextualized tool alone. That tool would be accepted or rejected based on its empirical worth to the recipient culture. In studying the transmission of acupuncture, I also had the opportunity to assess the value of these models.

From the first appearance of acupuncture in Europe in 1683 until its first decline in visibility in the mid-eighteenth century, needling offered a clear historical example of culturally saturated medical performance. Acupuncture entered the social realm of European medicine, but was unable to surmount the barriers to cross-cultural practice. The Far Eastern origins of acupuncture were a prominent part of discussions surrounding it; detractors of the technique disparaged the alien theories to which the needle was connected, while its supporters either translated those theories into western terms or simply ignored them. Acupuncture remained unassimilated, although it was not universally dismissed by the medical commentators. Moreover, in this first wave of transmission, acupuncture lost much of its complexity. In East Asia, therapeutic needling was performed to correct imbalances in the body's circulating vital energy, to remove blockages in the channels through which that energy – in the form of an imponderable fluid – flowed, or to treat disease in the organs which produced and processed the vital fluids. Needles were inserted into mapped points, located where the various energy channels were believed to come particularly close to the surface of the body. Each of these points was considered to have particular effects or potencies, relating to the channel or channels on which it lay, and affected by various environmental considerations. Despite the transmission of both acupuncture maps and the theories which explained them, European understandings of needling never included specific mapped loci, related in particular ways to particular diseases. The components of acupuncture had been transmitted separately – the needles and maps by a medical author writing to a medical audience, and the accompanying theory by both medical and lay observers addressing a general but elite audience. The two streams of information were not successfully reunited until the twentieth century.

After its fragmentation, the practice of acupuncture in Europe rapidly began to diverge from its Asian antecedent. By the early nineteenth century, as a second expansion of European and British acupuncture-use was rising to its peak, the technique had multiple forms, of which only acupuncture-analgesia bore any relation to the Chinese practice of acupuncture. However, the British surgeons and physicians who used the needle considered themselves to be using a technique rooted in Chinese medicine, though flowering only after extensive European (and particularly French) cultivation. I have followed their self-perceptions up to a point; I have also adopted the distinction made by several leading proponents of acupuncture between acupuncture which was intended to cure disease and relieve pain and those uses of the needle which had purely physical and local aims. However, in using participants' own definitions, it is essential to bear in mind the fact that 'acupuncture' in Britain was never practised, and only rarely discussed, in conjunction with the rules and conceptions of Chinese acupuncture.[24]

Almost immediately it becomes clear that the conventional constructions of technology as either culturally neutral or culturally saturated cannot adequately explain the acupuncture phenomenon. Certainly, a technique as complex as acupuncture could not be embodied in its entirety by the simple technology of the needle. Yet the flexibility of the term 'acupuncture' in western use – the way in which a range of needle-based medical practices were called acupuncture, despite their western origins and different therapeutic aims – indicates that the term came to be more closely associated with the tool than with the complex and culturally alien technique which the term was invented to describe. For the purposes of this study, I suggest that acupuncture be read as a palimpsest, inscribed at first with the rich theoretical and cultural structures of Chinese medicine; then over-written with the preoccupations of the physician-observers in East Asia. Scoured in the transmission process, acupuncture (and in some ways, the China which it and other exported artefacts came to represent) became an apparently clean sheet, upon which first European intellectuals and then French and British practitioners wrote their own definitions of 'acupuncture'. Under these layered erasures and re-inscriptions, the Chinese substrate was obscured; however, given the pattern of correspondence between perceptions of China and responses to acupuncture, it would be rash to assume that acupuncture's bond with China was obliterated.

1
Expectations and Expertise: Early British Responses to Chinese Medicine and Technology

> Every nation and tribe has what we may call its national thera-
> peutics and nosology. It has some conceptions of disease pecu-
> liar to itself, some modes of treatment not observed elsewhere.
> In principle and extent, they may be very humble, in detail
> united with error and mistake, but I think we should have to
> search a long time before we found one that would not afford one
> fact for our information, or one hint to awaken our curiosity.[1]
>
> G. T. Lay, 1841

In the 1840s with Britain teetering on the brink of empire, G. Tradescant
Lay's pleas for a sensitive and receptive approach to what were rapidly
becoming 'mere' colonial cultures fell on deaf ears. In his own day, Lay
was exceptional particularly for directing attention to the scientific and
specifically medical expertise of colonial peoples. In fact, his words
recalled a tradition of discovery-scholarship which was already in
decline when Britain sent its first official embassy to China in 1792.
This mission, led by Lord Macartney, was primarily aimed at enhancing
Britain's status and trading position in the Far East. Its members, while
eager to see as much of China and the Chinese as possible, did not
expect to learn much from Chinese culture – only to learn more about
it.[2] Their expectations reflected a more general shift in European self-
positioning with respect to other cultures. European travellers of
the sixteenth and seventeenth centuries had observed with interest
the inventions and expertise of the exotic peoples whom they encountered. Upon returning to Europe, they met with intellectually
receptive audiences, to whom they were able to present non-western

cultural productions and regimes as worthy of imitation. However, by the mid-eighteenth century, the tone and content of such accounts were rapidly changing, as were the demands of both audiences and governments; reports began to emphasize local products and raw materials, rather than indigenous knowledge and expertise. Although travellers continued to comment upon the learned cultures, technologies and medical practices they observed, a new category of knowledge had superseded those subjects in determining the hierarchy of civilizations: abstract 'scientific' understanding of workings of the natural world.[3]

In this chapter, I begin the process of exploring British responses to Chinese claims of expertise in medicine and science.[4] Encounters between members of the Macartney mission and their Chinese counterparts are central to this exploration: their accounts of China were hugely influential in shaping both British policy and popular views of China in the nineteenth century, largely because they represented Britain's first formal – and indeed 'scientific' – evaluation of the Celestial Empire.[5] The first section of this chapter describes and interprets medical interactions between the Britons and their Chinese interlocutors. In subsequent sections, I will examine the different lenses – imposed by professional commitments, by the changing structures of eighteenth-century western medicine, by class, and by established expectations – through which Chinese expert knowledge was observed. Each of these lenses was the product of a specific historical moment at the close of the eighteenth century, when rival models of medicine, of the body, and of appropriate social relations were in conflict with each other. Ideas of China, too, were fluctuating; the dazzling China presented in two centuries of Jesuit propaganda no longer shone so brightly in comparison with the West, yet the glowing after-image lingered and was much contested. Although the individuals of the Macartney mission were shaped by a shared culture constructed around nationality and language, they were equally products of the disjoint subcultures which proliferated in the atmosphere of flux. I will argue in this chapter that their different reactions to Chinese expertise reflect the debates, preconceptions, and ambitions current within each of these cultural microcosms with at least as much accuracy as they reflect China itself.

The Macartney mission: commerce and discovery

Nominally in honour of the Emperor QuianLong's eightieth birthday, the Macartney Embassy was actually directed to work for better trading conditions in China. Specifically, Britain sought the removal of various onerous regulations on trade in the open port city, Canton – described by one member of the mission as 'the Chinese knavery practiced at Canton'.[6] Reliable information about China was scarce. Indeed, the only Europeans officially sanctioned to learn Chinese were those who agreed never again to leave China – the Catholic missionaries. These priests had been sending information back to Europe for centuries, and the China they portrayed had permeated European culture, sparking the popular chinoiserie craze of the early eighteenth century, as well as informing the idealized Cathay of the *philosophes*. However, as sources, they were falling into disrepute. John Barrow's comments, published in his account of the Macartney mission, exemplified widespread British scepticism about the missionary and European descriptions of China: '[W]ith regard to China ... it may be considered as unbeaten ground by Britons ... we have not yet heard the sentiments of an Englishman at all acquainted with the manners, customs and character of the Chinese nation. The voluminous communications of the missionaries are by no means satisfactory.'[7] They were considered particularly unsatisfactory as sources of technical information, whether about Chinese manufacture or Chinese medicine. This distrust of the existing material on China, and the increasing importance of the China trade (especially for tea), ensured that a secondary goal of the mission was to observe and report on the Chinese interior, with its people, manufactures, sciences and customs – and of course its mysterious and all-powerful Emperor Quian-Long. To this end, Macartney included a natural philosopher and experimentalist, a surgeon, a physician and naturalist, and two botanic gardeners in his complement. With an entourage of artists, mechanics, soldiers and musicians, the mission totalled almost 100 persons. It was as much a continuation of the age of exploration as Britain's opening salvo in the siege of China.[8] And while an appraisal of Chinese medicine was desultorily included amongst the embassy's many goals, western medicine was explicitly enlisted as an engine in its moral and economic siege:

> In Doctor Gillan, the embassy was provided with a skillful physician; a circumstance desirable ... from the consideration that, after his arrival in China, the successful exercise of his profession among a people supposed to be far behind Europeans in every kind of science, might excite their admiration as well as gratitude; and thus contribute to the general purposes of the mission ...[9]

Ironically, given such assumptions of Chinese medico-scientific inferiority, one of the earliest encounters between the ambassadorial party and the people of China occurred over the sickbed of the aggressively sceptical John Barrow.

Encountering illness: lay expectations of Chinese medicine

After a tedious passage from Macao to the Chinese mainland, Lord Macartney and Captain Erasmus Gower sent the ship's tender, the *Clarence*, and a small party to shore at Chusan in search of native pilots who could guide the larger ships to Tientsin. George Staunton, the Embassy's historian and Macartney's right-hand man, led this group and recorded the subsequent events:

> During the stay of the *Clarence* in Chu-san harbour, one of the persons who came in her was seized with a violent cholera morbus, in consequence of eating too freely of some acid fruit he had found on shore. As no medical gentleman, nor any medicines happened to be on board, inquiries were made immediately for a Chinese physician to administer, at least, some momentary relief to the patient, then labouring under excruciating torments.[10]

Official British expectations of Chinese medicine, clearly, were not high, since 'momentary relief' was the most they hoped for even in this relatively simple case. In fact, in his own account of his illness, Barrow claimed that he had requested only that two drugs be sent, and had neither expected nor desired Chinese medical advice. In any case, a physician was swiftly provided by the Chinese authorities. As far as can be ascertained from the details given in Staunton's account, the physician treated Barrow according to traditional Chinese norms of practice, and apparently to good effect.[11] However, despite its happy outcome, neither patient nor chronicler was satisfied by this encounter with the prowess of Asian medicine. Staunton's *Authentic Account of the Embassy* describes the entire incident, emphasizing the diagnostic procedures:

> A physician soon arrived; who, without asking any questions about the symptoms or origin of the complaint, with great solemnity felt the pulse of the left arm of his patient . . . moving his hand for several minutes backwards and forwards along the wrist, as if upon the keys of a harpsichord . . . He remained the whole time silent, with eyes fixed, but not upon the patient, and acting as if he considered every distinct disease to be attended with a pulsation of the artery peculiar to itself, and distinguishable by an attentive practitioner.[12]

Where Staunton's tone (if not his attitude) was that of the objective reporter, Barrow reported his treatment with mocking scepticism:

> With a countenance as grave and a solemnity as settled, as was ever exhibited in a consultation over a doubtful case in London or Edinburgh, [the Chinese physician] fixed his eyes upon the ceiling, while he held my hand, beginning at the wrist, and proceeding towards the bending of the elbow, pressing sometimes hard with one finger, and then light with another... This performance continued about ten minutes in solemn silence, after which he let go my hand and pronounced my complaint to have arisen from eating something that had disagreed with the stomach.[13]

Staunton and Barrow were both educated – Staunton took an MA in medicine from Montpellier – and both could afford a comprehensive knowledge of the mores and manners of European physic (although Barrow, coming from a smallholding family, experienced elite medicine first in his patron's household, as tutor to Staunton's son). In each case, their disapproval stemmed largely from their disbelief of the alien diagnostic techniques used in Chinese medicine. Barrow described this strange behaviour explicitly as 'performance', and shared with Staunton the comparison of pulse-reading to the showy gestures of a harpsichordist. Certainly, the Chinese physician's gestures and attention to the pulse were much more elaborate than those of his British contemporaries.[14] The British participants judged the Chinese practitioner disingenuous, and their verdict on this high-profile case of cross-cultural practice was therefore discouraging, even though the patient regained his health: '[The Chinese physician] pronounced the present complaint to arise from the stomach, as indeed was obvious from the symptoms, *of which it is very probable he had information before he came*; and which soon yielded to appropriate medicines, supplied, at the patient's request, by him.'[15] Barrow concluded his retelling of the event by remarking that, 'I shall not take upon me to decide whether this conclusion was drawn from his skill in the pulse, or from a conjecture of the nature of the complaint from the medicines which had been demanded... or from a knowledge of the fact.'[16] In this example of a medical encounter, the exotic opacity of the Chinese medical performance rendered its apparent therapeutic efficacy even more incredible – at least to an audience whose expectations were firmly embedded in the protocols and praxis of western medicine.

Cultural specificity and diagnostic consensus

Staunton and Barrow's dismissal of an alien system of therapeutics and its concomitant assumptions about the body accords well with the models of medical community and therapeutic contingency developed by historians of medicine since the late 1970s.[17] Barrow and Staunton, as affluent consumers, were full and informed participants in the medical culture of eighteenth-century Britain. As such, they had strong views on the proper practice of medicine.[18] Both men responded with confident hostility to the visible attributes of Chinese medicine and in particular to Chinese pulse diagnostics. In much the same way that their descriptions of China expose their reactions to developments in European politics and trade, the specific traits of Chinese medicine which Barrow and Staunton found objectionable reveal their underlying attitudes towards contested trends in *western* medicine, as well as their expectations of Chinese practice. For example, their shared reaction to the status and role of the patient in the therapeutic encounter highlights one point of conflict between the Asian and European systems. Barrow and Staunton jointly mocked the idea that patients in China were mere bodily texts for their doctors' reading, rather than being themselves the readers and reporters of their own bodies, as was the case in the British practice of the day.[19] Similarly, both men expressed doubts about the honesty of the Chinese practitioner when he claimed that his correct diagnosis was based on the reading of the pulse alone.[20] Clearly these consumers did not accept the Chinese belief that the pulse revealed detailed information about all the organs of the body, even though the Chinese reading of this hypothetical compound pulse had confirmed, rather than conflicted with, the western diagnosis. Although both groups understood the pulse to convey reliable information about the body, the western observers considered that information to be at most two-dimensional, based on the speed and force of the heart-beat. This interpretation matched the newly established scientific conception of the pulse in western medicine, but was completely incompatible with the explanatory system which prevailed in China, where pulses taken at specific points on the arm were considered to reveal the state of different organs.[21]

The professional opinion: prepared for failure

In a case where both medical cultures produced similar diagnoses and remedies – Barrow noted that his self-prescribed medications (opium and rhubarb) 'met with [the] entire approbation' of his Chinese healer – British doubts were raised primarily by the devaluation of the patient's subjective experience of his illness in diagnosis. Their qualms reflected the hot debate in contemporary European medicine between the clinicians of the Paris School – who had begun to assert their own ability to independently read the patient's body – and their lay and professional opponents across the Channel, who argued for the centrality of patient accounts. In addition, the British observers' misgivings were strengthened by the discrepancy between European and Chinese interpretations of the pulse. The weight of disapprobation in both of these lay narratives rests on the exclusivity of Chinese reliance on the evidence of the pulse, and not on the disagreement between the expert opinions of Chinese and western physicians about how much the pulse could reveal about bodily states.[22] Barrow and Staunton filtered their encounter with Chinese medicine through their experiences of western medicine, but as non-practitioners, they were less constrained by existing professional assessments of Chinese practice. However, accounts of the Macartney mission also include a description of Chinese medicine written by a professional western physician observing and practising on Chinese patients.

Whether his expectations were formed by the travel books and medical commentaries of the mid-eighteenth century or by the more immediate experiences of his fellow-travellers, Hugh Gillan, the Embassy's physician, anticipated no more than hide-bound traditionalism from his Chinese counterparts. Indeed, he opens his 'Observations on the State of Medicine, Surgery and Chemistry in China' with a series of criticisms which, although clearly based on standard accounts of China rather than personal observations, indicate a highly selective and negative interpretation of those sources. Even DuHalde's glowing and wildly popular description of China expressed sharp disappointment with Chinese medical theory and knowledge of the body – and indeed all of China's scholarly chronicles of the natural world:

> When we cast our eyes on the great number of libraries in China, magnificently built, suitably adorn'd, and enrich'd with prodigious collections of books...one would be apt to believe, that of all the nations in the world, the Chinese must be the most ingenious and learned. However a small acquaintance with them will most swiftly undeceive one. 'Tis true, we must acknowledge that the Chinese have a great deal of wit: But then is it an inventive, searching, profound wit? They have made discoveries in all the sciences, but have not

brought to perfection any of those we call speculative, and which require subtilty and penetration.[23]

DuHalde certainly did not condemn Chinese medical practice, and indeed was impressed by the diagnostic pulse-reading of their finest practitioners ('I speak not here of those quacks who profess the art merely to get a livelihood, without either study or experience: But of the skilled physicians, who, it is certain, have acquired a very extra-ordinary and surprising knowledge in this matter...').[24] However, by the 1730s, anatomy was emerging as a central guarantor of medical credibility in Europe. DuHalde included a detailed description of Chinese conceptions of the body in his *Description*, but he cautioned his readers:

> It cannot be said that Medicine has been neglected by the Chinese... But as they were very little versed in natural philosophy, and not at all in anatomy, so that they scarcely knew the uses of the parts of the human body, and consequently were unacquainted with the causes of distempers, depending on a doubtful system of the structure of the human frame, it is no wonder they have not made the same progress in this science as our physicians in Europe.[25]

In Britain – with its established distaste for scholasticism and its new fondness for dissection – medical professionals were even more critical of Chinese medical knowledge. Drawing upon the scholarly texts produced by medical men posted in Asia as well as general works on China and Japan, one medical commentator, Dr R. James, described Chinese practice as composed essentially of obscure texts and straightforward simples. This version of Asian medical practice noted the points which so surprised Staunton and Barrow in Chusan, but with a difference in emphasis:

> They acquire a knowledge of diseases by a long and tedious observation of the pulses... Besides the pulse, they consider the eyes, the tongue, and the face of the patient, but neglect all other circumstances from which the prognostics might be drawn; for *they neither interrogate the patient with respect to his state, nor inspect the urine of those who are under their care... After this they have recourse to a most antient [sic] book, which is the standard of their practice, find out the denomination of the pulse, and the remedies appropriated to the particular disease, of which it is the concomitant symptom.* Most of their medicines are simple and easily prepared, such as decoctions.[26]

The idea of treating all patients' ills through the use of general tables in an ancient book would certainly have seemed shocking and ludicrous to British eyes; after all, the medicine – and especially the pharmacy – of the day was rooted in a notion of the particularity of individual cases.[27] James continued by detailing the more specific failures of Chinese physicians, especially their ignorance of chemistry and anatomy and the fact that they 'never admit of phlebotomy'.[28] But most importantly, James ridiculed Chinese theories and medical knowledge, and the positive tone in which both had previously been reported:

> [The Chinese] imagine, that there is a certain circulation of the blood and spirits, which conveys the radical moisture, and the native heat, thro' the veins and vessels of the twelve members... This phantastic and ridiculous account of a circulation of the fluids in a human body has induced some of the less wary and circumspect of the *Europeans* to assert, that the circulation of the blood was very early known to the *Chinese*... They have formed a pompous kind of pathology in order to account for painful and spasmodic disorders... Their theory in general, however antient [sic], is yet very imperfect and unphilosophical...[29]

The assumptions apparent in James's 1743 essay, and especially his judgement that explanations of the body which put theory prior to anatomy were 'pompous', were quite common in the medical literature of the mid-century. Moreover, Gillan received his medical training at the University of Edinburgh, in 1786 – in other words, just after the popularity of Brunonianism, with its monistic pathology and universal therapeutic of stimulation had peaked, and while the University establishment's reaction was at its fiercest.[30] Edinburgh had been, in any case, the first British university to wholeheartedly embrace medicine's collateral sciences in its teaching. At the University of Edinburgh, systematic and rational medicine was the hallowed goal, and purely empirical practice 'the most contemptible of medical philosophies'; experience played a crucial role in good medical practice, but only when guided and constrained by science and system.[31] Thus Gillan was unlikely to appreciate the one virtue conceded to Chinese medicine by even its harshest critics: that its practitioners, through long experience and careful observation, equalled their western counterparts at least in the empirics of practice. Textual evidence indicates that DuHalde's *Description* was included in the Embassy's travelling library, and certainly DuHalde's voice echoes in Gillan's first-hand account of medical practice in China. But Gillan's report shows none of DuHalde's esteem for medicine as an art, or for practical experience as a source of medical authority; rather, he complains throughout about the 'unphilosophical' and 'unsystematic' nature of Chinese medical thought.

On his arrival in Jehol, Gillan had the opportunity to meet with Court physicians, hear their interpretation of a case, and explain – to the highly influential patient, as well as his doctors – the western alternative. The patient who called in this rarity, the European doctor, was Ho-Shen (or Heshen; also referred to by the title of his post, 'Colao' or 'Grand Secretary'), a Manchu courtier and the Emperor's long-standing favourite. 'He sent . . . to the Embassador, a request to send to him his English physician, whom he wished to consult upon his case'; Gillan accompanied the messenger to the Colao's house, where he was met by 'some of the principal persons of the [medical] faculty then at court'.[32] Ho-Shen's illness, as he and the Chinese faculty explained, involved both shifting pains and inflammation in his joints, back, and loins, and 'excruciating pain about the lower part of the abdomen' accompanied by swelling in the groin.[33]

Two reports of this medical consultation exist; the first, Dr Gillan's notes on Ho-Shen's case and on several other cross-cultural medical encounters, only circulated privately. These notes formed the basis of Staunton's published account. The Staunton version was one of the most widely available early descriptions of Chinese praxis in English, casting a visible shadow upon future British responses to information about Chinese medicine well into the nineteenth century. In both renderings of this meeting, Chinese theory was described in greater detail than in the stories of Barrow's illness. Moreover, in this case, where European and Asian diagnoses differed substantially, Chinese medicine was presented as empirically inadequate, as well as theoretically risible. Staunton's edited extract of Gillan's notes emphasized the stranger aspects of Chinese medical practice, opening with a portrayal of the different method each culture used to determine Ho-Shen's symptoms and define his illness:

> These circumstances [Ho-Shen's symptoms] the Doctor learned from the Colao himself; who, however, was surprised at such a number of questions, which the other physicians had not thought it necessary to make. They drew their indications chiefly from the state of the pulse, in the knowledge of which they boasted the highest skill. According to their ideas, every part of the body has a pulse particular to itself, which indicates what part of the system suffers.

They considered the pulse as a general interpreter of animal life... and that, by its means alone, the nature as well as seat and cause of disease, could be ascertained without the necessity of any other information relative to the patient.[34]

Gillan's notes here differ subtly from the closing lines of Staunton's summary of them.[35] While the impression given in Staunton's text is that the Colao's case was in fact handled in this way, Gillan was actually discussing the typical claims made by Chinese healers, rather than the behaviour of the physicians in any particular case – his narrative speaks more of bedside arrogance than ignorance.

In general, Gillan was more aware of the centrality of medical protocol and patient demands in the doctor–patient interaction, and implied that the reticence of Chinese physicians might be a matter of fulfilling patient expectations rather than simply a display of professional arrogance: '[Ho Shen] seemed a good deal surprised at my asking him so many particular questions, *which is not customary for physicians in China to do in any case.*'[36] Gillan himself (as Staunton noted) made considerable efforts to make his prestigious patient feel at ease:

After the first ceremonies were over... he made me sit down on the couch... beside him and presented me first his right arm, and next his left... believing that from their indications alone I could tell him everything respecting the nature, cause, and actual state of his complaints. In compliance with the customs and prejudices of his country, and that I might not at first shock himself or his physicians, who stood around us, I felt with seeming attention and gravity all his pulses on both arms and continued to do so for a long time.[37]

Gillan's account also points out areas of consensus between Chinese and European medical theory. Most importantly, he expressed a far more sophisticated understanding of the pulse as a vehicle for information about the body, and a more exact description of what made this view incompatible with the European understanding of the pulse:

There is certainly a very close connection established by Nature between the state of the pulse and the general affection of the system, with regard to frequency, force, fullness, hardness and softness, etc., by which we are enabled to form our judgements respecting the nature and cure of many diseases. But the Chinese physicians carry their pretensions far beyond this. According to their ideas, every viscus and every part of the body has a particular pulse belonging to itself, which indicates with certainty what part of the system suffers, and how it is affected in disease.[38]

The European medical profession's construction of the pulse as a diagnostic tool was clearly more nuanced than was apparent to lay observers like Staunton and Barrow. Even at the end of the eighteenth century, it retained some of the plastic importance granted it by early users.[39] However, Gillan's professional opinion of the Chinese multiple pulse, while expressed more moderately and drawing on more information than Staunton's reading of it, was nonetheless quite negative: '[I]t is evident from the nature and structure of the human frame, as well as from the circulation of the blood, that the general doctrines of the Chinese physicians upon [the pulse] must be false. Nothing indeed but the grossest ignorance of anatomy and physiology could make them believe such absurdities.'[40] It is indicative that Gillan drew his authority explicitly from those western bastions, the collateral sciences, rather than from assertions of therapeutic superiority. Indeed, in his attempt to convince Ho-Shen himself that the western understanding of the pulse was the correct one, Gillan relied on the demonstrable accuracy of the scientific view rather than on any gains in medical efficacy derived from it. He explained the western theory of the pulse, and then asked Ho-Shen to perform a small experiment on his own pulses:

> I told him that we were not accustomed to examine the pulse in various parts of the body in Europe, that we usually felt it in one place only, because we know that all the pulses correspond together and communicated with the heart and other parts by means of the circulation of the blood, so that from knowing the state of one ... we knew the state of the rest. He seemed to listen to me with astonishment ... At my request, and to satisfy himself on this subject, he applied the forefinger of one hand to the left temporal artery and the same finger of his left hand to his right ankle, and to his great surprise he found the beats of his pulses simultaneous.[41]

Gillan's notes reveal that he explained the 'variety of questions respecting internal sensations and external circumstances' underpinning western diagnosis (and sought permission to ask such questions) only *after* his experiment's successful completion. As Gillan (and Staunton) presented this event, experiment served not only as the solid and freely exchangeable currency of proof, but also 'bought' sufficient authority for Gillan to deviate from China's established clinical forms and ask his diagnostic questions. The experiment also provided an opportunity to present the Chinese as possessed of a simple, static and dogmatic medical culture. Thus Staunton's account noted with satisfaction the surprise, astonishment and, on the part of the doctors, embarrassment, attributed to the Chinese observing and participating in it.

In both the published and unpublished reports, the authors made it clear that Chinese theories about the pulse led inevitably to an incorrect diagnosis. Such a diagnosis could only produce a ridiculous therapeutic strategy. In Ho-Shen's case, Gillan and Staunton described both the observed process of diagnosis and the unseen therapy in highly negative terms:

> In consequence of this opinion of the nature and cause of the disease, the method of cure was to expel the vapour or spirit immediately... The operation had been frequently performed... with exquisite pain to the patient... [T]he disease continued in its usual course: but this, from the authority and information of his pulses, was entirely owing to the obstinacy of the vapour... In their treatment of this disorder, the physicians had exhausted all their skill to no purpose.[42]

In this passage, Staunton was ostensibly describing the specific practice and rationale followed by the Chinese physicians. He also conveyed the far more damaging impression that they diagnosed Ho-Shen according to their dogmas about the pulse rather than observation: 'After a full examination of the Colao's pulses, they had early decided...' on what was portrayed as an explanation of convenience. Gillan noted incredulously that 'the whole of his complaints and all their change of place and symptoms' were attributed to one cause, 'a malignant vapour or spirit... which shifted itself from place to place', which was itself unknown to and incompatible with current western medical knowledge. His tone suggests that he considered his Chinese colleagues to have simply invented a mechanism capable of explaining all of the multifarious symptoms. Rhetorically too, this was the strongest blow to the credibility of the Chinese faculty. The idea that Ho-Shen's two major symptoms – joint aches and abdominal spasms – could both be explained by a shifting attack of gas would have astounded the western medical community, consumers and providers alike.[43] The unities which were immediately apparent to the Chinese faculty, viewing the body as a system whose delicate balance depended on the flow of *qi* – and therefore viewing the symptoms as inevitably 'part of the same disease' of blocked or inappropriate *qi* – were invisible to western eyes. In addition to being utterly unconnected in the new anatomy-based nosology, orthodox British practice treated the former ailment – diagnosed as rheumatism – medically and systemically, and the latter – reported to be a hernia – surgically and specifically. Thus, the diseases were properly the province of different professional groups, and considered in different ways by western medical professionals. As Gillan's report of his questions also illustrated, the two ailments were not even seen to depend on similar environmental or emotional states; that the Chinese faculty could have missed an indicator of separate causation obvious to even the least sophisticated western layman would have seemed truly barbaric to a British reader.

Staunton repeated Gillan's description of the proposed cause of Ho-Shen's symptoms, retaining his use of terms which to British ears would inevitably connote anachronistic superstitions, or at best a species of coarse ignorance about the body. His language suggested that either the Chinese attributed disease to malignant spirits or to a sort of shifting, self-generating flatus. Elsewhere, Staunton was even more explicit about his view that the Chinese relied on animism and superstitions in their medicine (and suggested a root cause for that reliance):

> There are in China no professors of the sciences connected with medicine. The human body is never, unless privately, dissected there. Books, indeed, with drawings of its internal structure, are sometimes published; but these are extremely imperfect; and consulted, perhaps, oftener to find out the name of the spirit under whose protection each particular part is placed, than for observing its form and situation.[44]

Staunton clearly found the association between an absence of anatomical knowledge and a dependence on animism both immediate and convincing.

Chinese stories: narratives of contention and narratives of change

> The meaning of traditional therapeutics must be sought within a particular cultural context; ... To understand therapeutics ... its would-be-historian must see that it relates on the one hand to a cognitive system of explanation, and on the other, to a

patterned interaction between doctor and patient...which evolved over centuries into a conventionalized social ritual.[45]

Rosenberg, 1985

Clearly, these stories, of Chinese medicine practised on British subjects and of British medicine practised on Chinese subjects, reveal a rich network of embedded assumptions about proper medical practice and understandings of the body. However, they also reflect a fluid and transitional stage in European medicine, especially through the particular sites of contention between European and Asian practice. The three narrators, professional and lay members of the medical community, reject or challenge the same parts of Chinese medicine. Specifically, they resist the idea that the patient can be divorced from the illness, and that the patient's unique perceptions of his or her own body can be safely discarded from the diagnostic repertoire. In relation to the pulse, the only physical sign which the professionals of both cultures accept unanimously, all three western narrators reject the wide-ranging diagnostic authority given it by the Chinese, and strip the pulse of its more subtle nuances. Strikingly, the evidence upon which Gillan's rejection is explicitly based is 'scientific' and not drawn from an assessment of Chinese practice, while Staunton's rejection leans more heavily on its diagnostic failure than on its incompatibility with the new sciences of anatomy and physiology. The British observers agreed that no single sign could reveal so much, and that certainly the Chinese physicians could not legitimately claim so extensive an authority over the body. Their negative response to Chinese medicine was moulded by the new, largely anatomical, understanding of the body produced during a century of medical exploration in Europe; however, it was also inflected by the increasing tensions surrounding trade, and mediated by their personal disappointment with the scant progress made by the much-vaunted Embassy. Disdain was not the inevitable response of one culture to the medicine of another; indeed, Sir John Floyer, an early British advocate of sphygnometry, used the Chinese tradition of pulse-diagnostics as a source of evidence and authority.[46]

The power of the pulse: an early response to Chinese diagnostics

European medicine entered a period of radical change, in terms of both its intellectual and its professional values during the second half of the eighteenth century. The effect of these changes on interpretations of

Chinese medicine can best be judged by comparing early responses to Asian practice with those of the Macartney Embassy. Sir John Floyer (1649–1734) was Oxford-educated in both arts and medicine; received a knighthood, probably for services in the intrigues of James II; advised Samuel Johnson about his health; and was recognized as a medical innovator and modernizer in his own time.[47] Writing in 1701, Floyer responded to newly available information about Chinese medicine with enthusiasm, although not without criticism. Significantly, his intellectual and professional response to this culturally alien material was to assimilate the Asian medical traditions, re-interpreting them in the language of western medicine. Indeed, and in marked contrast to James, writing only four decades later, Floyer placed Chinese medicine on the same level as the Galenic corpus, reinterpreting both of these older traditions of medicine 'according to the new Anatomy, and our present philosophy'.[48] Even the title-page summaries of his *The Physician's Pulse-Watch* clearly expressed his opinion that these two traditions, while separate, were of equal value and relevance. The first part of his treatise was described as 'I. The Old Galenic Art of Feeling the Pulse is describ'd, and many of its Errors corrected', while the third section came under the heading: 'III. The *Chinese* Art of Feeling the Pulse is describ'd; and the Imitation of their Practice of Physick, which is grounded on the Observation of the Pulse, is recommended.'[49] If anything, Floyer rated Chinese practice above that of the Galenic tradition. His judgement, based on a wide range of reports from both China and Japan was that, 'as to physick and Chyrurgery they are Experts, and their rules of the Art differ not much from those of the European Physicians; for at first they feel the pulse like them, and are skillful in discovering by the same the inward distempers of the body'.[50]

Floyer tried to explain and synthesize the Chinese and western accounts; for example, he took the Chinese description of individual pulses as revealing the state of particular organs seriously, but not literally. To explicate it to his medical audience, he observed that each of the organs named by the Chinese produced a particular fluid, and declared that the Chinese were speaking allusively in naming, say, the liver, rather than the bile which that viscera produced:

> 'Tis ridiculous to believe that the pulse can depend in its alterations on the solid parts of any viscera, but it does evidently alter by the fluids; therefore, 'tis obvious that the Chinese respect the fluids, which are secreted by those parts in feeling of the pulse; and if this

be a fair conjecture, I have probably accommodated the Chinese and Grecian art of feeling the pulse.[51]

Floyer quoted from a translated Chinese medical text, the Nuy Kim, line by line, giving what he considered to be a mundane and down-to-earth paraphrase in western terms as he went along; thus 'the eyes are the windows of the liver' was taken to mean simply that diseases of the liver presented symptoms in the eyes – Floyer cited jaundice as an example here. Unlike his British successors, Floyer did not *expect* Chinese medicine to be ridiculous, and therefore when its assertions seemed to his ears laughable, he assumed that the error was in the translation, rather than in the ideas originally expressed. This is not to say that he did not hold any stereotypical views about the differences between western and Chinese abilities; although his receptivity to innovations from both East and West was remarkable even among his contemporaries, Floyer was no such anachronism. His report on a text by Andreas Cleyer on Chinese medicine illustrates both his openness and its limits:

> In this discourse quoted from Cleyer, I find good sense, tho' express'd in the Asiatic way, whose words are sorts of hieroglyphicks, as well as their characters; and their expressions are fitter for poetry and Oratory, than phylosophy; the Asiatic have a gay luxurious imagination, but the Europeans excel in reasoning and judgment, and clearness of expression.[52]

In fact, Floyer used the Chinese science of the pulse to provide historical roots and justifications for his own innovation, the 'physician's pulse-watch'.

> After by my pulse watch I had found the most healthful pulses, I easily discern'd what were the exceeding and the deficient pulses. 'Twas easie for me to take Indications from the hot or cold pulses... After I had reflected on what I had done I found my notions hit with the *Chinese* Practice, about which I consulted many printed Travels...[53]

As this passage suggests, almost every description of China available in Floyer's time contained some mention of pulse medicine; it was the subject of intense curiosity particularly in this period, following the European discovery of the circulation of the blood. For Floyer, the ancient use of the pulse as a diagnostic tool, and the great efficacy attributed to Chinese medical practice by the Jesuits validated the idea of relying on a physical sign, while the intractable source of Chinese expertise – long experience, rather than any particular depth of anatomical knowledge – offered a reason for turning to a mechanical aide-de-camp, in the form of a pulse-watch and a set of tables. Compare this reasoning to the scorn James heaped on the use of similar diagnostic tables by the Chinese in the 1740s!

One reason for the great shift in attitude between Floyer's response to Chinese medicine and Gillan's was that in 1701 there remained a parity, within the profession, between the value of empirical experience and that of scientific knowledge. The idea of medicine as 'art' had certainly not been erased by 1792, but the notion that a medical science might be preferable was increasingly popular, especially as its potential benefits to the medical profession became more and more obvious. It did not diminish Floyer's own authority to assert that,

> Tho' neither the *Greeks* nor the *Chinese* knew the true Fabrick of the Organs of the Pulse, nor their true action and uses, nor the circulation of humours, and the causes of it, yet the *Greeks* discovered the polses of all diseases and humours, and passions: And the *Chinese* their Art of Physic on the pulse and its differences...[54]

simply because his credibility was not maintained by his knowledge of the collateral sciences of medicine, but by his success at the bedside and his elite Oxbridge education. The value placed on experience as a form of medical authority is even more evident in Floyer's conclusions about the Chinese. He insisted that translations of Chinese medical texts would be valuable to western medicine:

> 'Tis certain, their experience of this practice for 4000 years is much to be valu'd, because they are an ingenious nation; but we have at present an obscure account of it from the missionaries, who know nothing of this art; neither can they dexterously distinguish and separate the Chinese notions from matters of fact, and the real phenomena, to which all hypotheses are adapted, tho' they be very absurd.[55]

Floyer argued that this task had to be done by a physician who himself 'had a full experience of the Chinese method of practice, that he may more clearly describe their pulses, and accommodate them to the names we use'. Clearly, the assumption is one of an underlying unity, of at least a potentially shared conception of medical practice and the body. All that remained to be overcome was '[t]he ignorance of the Europeans in the Sphugmatic science, together with the hieroglyphick mode of the Chinese notions'. Intriguingly, Floyer both acknowledges the usefulness of medical science, and mildly scoffs at it. So he notes that, 'the want of anatomy does make [the Chinese] art very obscure, and gives occasion to use phantastical notions' but balances this flaw against the fact that 'their absurd notions are adjusted to the real phaenomena, and their art is grounded upon curious experience, examined and approved for four thousand years'.[56] He rebukes critics and doubters who dismissed Chinese techniques simply because of diagnostic terms like 'flying ribband' with the tart comment that,

> the Chinese distinguish the pulses by comparing them to something that feels like them; and they who will know their meaning, must discern the same by long experience in feeling the pulses; for they do not consider the pulse as geometricians do by its dimensions.[57]

Floyer was looking far ahead as he responded to the deluge of new medical information, whether from China or from the new perspectives offered by anatomy and chemistry. In plotting medicine's course towards professional status and respect, he was equally willing to incorporate Chinese medical insights as to critique their customs and lavish self-confidence. In particular, he was impressed by the possibilities of controlling the evidence of the body, without depending on the patient to access and assess it. After a lengthy discussion of how to interpret different pulse rates in individuals of different status, sex, age and climates, Floyer proposed the Asian model of doctor–patient interactions as the way to raise the status of the British medical profession: 'All these things must be considered if we design to imitate the Chinese skill, whereby we design to find out a disease, without being told of the symptoms; by which we may procure a great reputation among the vulgar.'[58] However, his proposal ill-suited the nature of the client-driven medical profession in the eighteenth century, and the attainable levels of accuracy in pulse diagnosis did not offer enough leverage to shift the balance of power towards the medical practitioner. The externally legible body, with its freight of authority and exclusive knowledge, was eventually produced over the course of the next 150 years, and had all the benefits for the medical profession that Floyer predicted, but it was never based on the multi-faceted pulse of Chinese medicine. Meanwhile, shifts in the intellectual culture of western Europe – and especially the rising authority of the sciences, as illustrated by DuHalde and James – filtered contentiously but irresistibly into medicine. Where humoral medicine depended on universally available and recognizable experiences of the body, 'scientific' medicine relied on perceptions of the body mediated by (and only available through) the novel ritual of dissection. As orthodox medical professionals struggled to gain sole authority over the body, many seized upon the exclusive knowledge offered by anatomy to ring-fence the all-too-accessible authority of experience. Medical faculties at Paris and Edinburgh were pioneers of this highly contested, but ultimately successful strategy. This shift in criteria, and especially the rigidity which characterized the new medical knowledge, effectively caused the points of difference between the medical cultures and languages of East and West to become irreducible, closed to interpretive malleability.[59] The cultural borders were closing, at least for the elites of China and Europe.

Through the eye of the needle: medical politics and the Gillan report

Broadly, the intellectual roots of these conflicts can be located in the widening divisions between experience and science, between shared and exclusive knowledge, and between natural philosophy and medicine in the West. However, 'scientific' medicine gained ascendancy only in stages, and the case of Ho-Shen also reveals the continuing tradition of systemic relationship between the health of the body and the state of its environment in western medicine. Consider what were seen to be appropriate diagnostic procedures, as illustrated by Gillan's interview with the Colao. The central argument he made in his report was the importance of questioning the patient directly. The answers he recorded from this direct questioning do illuminate the importance of establishing the exact physical symptoms experienced by the sufferer, but they also demonstrate an exploration of the potential connections between those physical signs and their natural and social context. The seasons in which specific symptoms occurred (and whether they followed a seasonal pattern), their frequency, location and duration were all a part of the diagnosis, as was the more general history of the patient in relation to episodes of illness. For Gillan, the inadequacy of the unitary Chinese explanation of Ho-Shen's illness was proven by this history:

> Upon full investigation, it appeared that he laboured under two distinct complaints. The first was rheumatism, which first attacked

him in the mountains of Tartary where he had long been exposed to cold and rainy weather, previous to the accession of his complaint . . . At that same time, he was confined to his tent for a whole month [with the symptoms] . . . The pain, swelling, and other symptoms recurred however, the succeeding spring, in cold moist weather, and had continued to recur . . . When I asked him more particularly respecting the pain and swelling of the lower part of the abdomen, he confessed he had from his infancy had some little swelling . . . It had, however, never given him any pain or uneasiness till about eight years ago when it suddenly increased to a very large size when he was making an exertion to mount a very tall horse . . . I examined the part and found a completely formed hernia.[60]

This older tradition, as was apparent in Floyer's comments on the Chinese pulse, had strong affinities to the Chinese ideas about the causes of health and disease which were incorporated into pulse diagnosis. However, by the time Gillan witnessed Chinese diagnostics in Jehol, the pulse had come to have a fixed place in western medicine as a monochromatic indicator of bodily health; any ideas about the body which were embedded in it by Chinese medical theory remained unsought and thus invisible. Here, Gillan's Scottish training as a physician probably distorted his observations.

In general, the theoretical system underlying Chinese treatment seems to have been hidden from Gillan (despite DuHalde's lengthy disquisition on the subject) perhaps because it was persistently visible only where it was not expected, or in shapes which were not recognizable to western eyes. He remarked at one point that, 'The use of general remedies, or such as act upon the whole system is equally unknown to them in medicine and surgery. In both cases, they think only of topical applications and medicine and seem not at all to conceive the necessity or advantages of general ones. They never bleed in any case.'[61] He then described the prescription of specific diets and exercises as a part of their treatment for venereal diseases, without noticing any contradiction. These restrictive diets surely were indicative of a systemic approach – certainly, Gillan did not argue that the Chinese believed that vinegar, wine or oil caused the lues venera. But diets clearly did not fit Gillan's idea of a remedy, or his model for systemic therapy, probably because the archetypal systemic remedy was blood-letting, with purging and vomits close behind.

Gillan's emphatic belief in the importance of the collateral sciences to medical practice – dogma among the University of Edinburgh medical faculty – is evident throughout his report on Chinese medicine. The document opened with a starkly critical assessment of Chinese expertise:

> The state of physic, both as a science and as a profession is extremely low in China; as a science, indeed, it can hardly be said to exist among them . . . They are totally ignorant of the anatomy and physiology of the human body, which is never dissected in China; nor do they seem to have any idea that such knowledge could be of any use to them in the treatment and cure of diseases. Their pathology and therapeutics must of course be extremely deficient and are for the most part erroneous. Natural history, natural philosophy, and chemistry, as sciences, are equally unknown to them. Their materia medica in consequence of this is extremely limited . . . Hence physic in China must be vague and uncertain in its principles and practice, and the profession itself necessarily becomes low and obscure.[62]

The absence of anatomy and physiology as recognizable sciences is portrayed as the cause of medical inadequacy, while the lack of chemistry and natural history made a weak pharmacopoeia inevitable – and thus blinded Gillan to the extensive, and much-reported Chinese materia medica. Gillan's comment on the weakness of Chinese surgery was even more definite about the paramount importance of anatomical knowledge:

> Surgery is indeed in a state still lower than the practice of physic among them . . . It has always been found in every age and in every country that the progress of surgery depended on that of anatomy, without an accurate knowledge of which it is impossible for anyone ever to excel as a surgeon. But in China, where anatomy is altogether unpractised and unknown, it is particularly remarkable how ignorant they seemed to be of every kind of surgical operation.[63]

Yet the centrality of anatomy to good *European* medical practice was still a contentious and highly political claim. Gillan was using the example of China to bolster his particular views on the appropriate form of training and knowledge for medical practitioners – views which while orthodox in Edinburgh, would have been contested by the academic physicians at Oxford and Cambridge and by many London doctors and surgeons.[64] Similarly, when Gillan condemned Chinese medicine on the grounds that 'no public schools or teachers of medicine, no professors of the sciences [are] connected with it', and asserted that Chinese profession's dependence on apprenticeships for medical training doomed it to disrepute, he was also making a strong claim about the situation in British medicine, where the superiority of college-trained over apprenticed medical practitioners was still hotly disputed.[65] When Gillan's observations of Chinese medicine yielded differences between Chinese practice and its European analogue which could be used as evidence supporting his positions within domestic medical debates, those differences became essentially non-negotiable. Given his strong support for organized medical education and the collateral sciences – and his desire to answer conservative challenges to their growing prominence – Gillan was very unlikely to report, or even observe, the contradictory signs of a medical, and especially pharmaceutical culture flourishing in China in the absence of anatomy and university training.[66] For example, Gillan described Chinese medicines (constructed without the benefits of chemistry) as 'very simple... They consist chiefly of vegetable powders, decoctions and a few extracts. These they seldom mix or combine, but for the most part exhibit them in their simple form.'[67] Meanwhile, Staunton watched a Chinese doctor prescribing for dysentery and mockingly described a Chinese prescription with many different elements.[68]

Gillan reported that Ho-Shen, after being examined, witnessing the pulse experiment and hearing the western diagnosis of his condition, 'desired the interpreter to tell me that my ideas and all that I had said were so extraordinary that it appeared to them as if it had come from an inhabitant of another planet'.[69] This statement was remarkable because it so eloquently expressed the sense of unbridgeable distance between the two therapeutic systems, without asserting that either was superior. However, Staunton reported this remark to the British public with a small alteration; it is this alteration which reveals the different stances of these two elite western observers:

> The Colao desired the Doctor's explanation of the nature of his ailments together with the methods of relief and cure which he proposed to be put down in writing; He... was pleased to say, *that his ideas appeared clear and rational*, tho' they were so new and distant from the notions prevalent in Asia, that they seemed as if they came from the inhabitant of another planet.[70]

Staunton replaced the original statement's tone of puzzlement with a humble admission of superior western rationality. Ideas, and especially ideas about technology, science and medicine, were to be exported from

Europe; their appreciation abroad was a mark of potential. Ho-Shen was portrayed throughout as an exceptional Chinese, precisely because he appeared more open to western ideas (and of course British trade) than his counterparts.

Anderson and Eades – perceptions of China below the salt

Given both modern insights into the cultural specificity of medicine and the hostile reactions of the Embassy's 'Gentlemen' to Chinese expertise, it is the more surprising to read Aeneas Anderson's report of a medical encounter very similar to the Barrow case. Anderson accompanied the Embassy as Macartney's valet, and wrote and published his version of events soon after his return.[71] His account of China resembled those of Barrow and Staunton in most respects, although it was in general more critical of the ways in which the British presented themselves to the Chinese. However, his opinion of Chinese medicine differed significantly from those of either Staunton or Gillan:

> Of the knowledge of medicine among the Chinese, I can say no more, than that I was a witness, in one instance, to a skillful application of it, in the case of John Stewart...who, on our return from Jehol, had been seized with the dysentery, which increased so much on the road, that at Waunchoyeng, there were no hopes entertained of his being able to leave that place...a Chinese physician was called to his assistance;...the man's case was explained to him by Mr. Plumb [one of the mission's Chinese Catholic translators]... The physician remained a considerable time with his patient, and sent him a medicine, which removed the complaint and restored him to health.[72]

Staunton described this same episode of illness, and the methods with which it was treated by the native practitioner; his report, however, contrasted sharply with Anderson's in both tone and detail:

> Another person belonging to one of the Embassador's suite, labouring under a dysentery, stopped at a Chinese inn, and was induced to consult a physician of the place, who, to the doctrine of the pulse, added a discourse upon the different temperaments of the human frame, and unluckily attributing his patient's suffering to the predominance of cold humours, prescribed for him strong doses of pepper, cardamoms, and ginger, taken in hot show-choo, or distilled

spirit; a medicine which so exasperated all the symptoms of his disorder, that he had much difficulty to escape alive to Pekin.[73]

Staunton's story uses the language of the British medical profession, and he clearly assessed the Chinese therapy from that perspective. The changes which had occurred in European medical theory between Floyer's time and his own are particularly evident in his contempt for the Chinese doctor's humoral diagnosis – a diagnosis which would have been completely intelligible and unexceptional to a British doctor at the beginning of the century, and which certainly elicited no objections from Anderson. The historical record notes only that the sufferer survived both his illness and the medicine which he received for it; no external evidence exists to validate either version of events.

The difference between these two accounts is unexpectedly large; there is disagreement even about the observed effects of the treatment. Moreover, the men who witnessed and reported this medical incident used markedly different criteria to evaluate it. In contrast to his employers, who judged Chinese medicine according to its fit with an ideal western model, Anderson based his assessment of the Chinese physician's authority – and by extension, the credibility of Chinese medicine – on what he saw as the objective success of the novel treatment provided. Thus, ironically, Staunton's gloomy prognosis may have enhanced Anderson's appreciation of the Chinese doctor's cure. Of course, it is possible that Anderson drew no distinction between Chinese and British medical practitioners. However, his inclusion of this event (barely mentioned even in Staunton's prolix account) and his note that the Chinese physician 'remained a considerable time with his patient', suggests that he was interested enough to actively observe and remember the encounter. The specific expectations of the members of the Embassy certainly would have depended on their degree of access to (and their respective sources of) information not just about Chinese, but also about British medicine – and indeed, about China itself. Given the costs of the available texts on these subjects, the class and education of Embassy members significantly affected the expectations which they brought with them to China.

This is not to imply that only the upper echelons had access to information about China, or expectations of that nation; however, the narratives of China to which non-elites did have access, and the expectations which these stories instilled, were quite different. Henry Eades, described by his companions and superiors as 'an ingenious and skilful artist in brass and other metals', was hired by the East India Company as a mechanic to assemble and if necessary repair the complex mechanical gifts brought for the Emperor. Originally from the Midlands, Eades was an established London artisan by 1792, and was neither educated, nor young, nor healthy. Nonetheless, he applied to go with the Embassy on the long and dangerous mission to China – and he died there, one day's journey away from Beijing. Staunton's brief eulogy, a paragraph in the *Account*, sheds a rare light on working-class perceptions of China in this early period. As Staunton noted, Eades was knowingly taking a substantial risk, both in business and health terms, by putting himself forward for the mission. This was no romantic turn on Eades's part; his reasons shine through Staunton's turgid prose, solid brass:

> [Eades] had conceived a notion that many improvements in the arts were practised at Pekin, which were little known in Europe; among others, that of making a kind of tinsel that did not tarnish, or at least that kept without tarnishing much longer, than any that was made according to European methods. He fancied that were he acquainted with such improvements, he should be enabled to provide handsomely for his family...[74]

Clearly, Eades expected to find valuable technical expertise in China, and the specificity of his expectations indicates a fairly well-informed source.[75] His hopes resembled more closely those of much earlier travellers and their audiences than those of his elite contemporaries. Eades went to China to see a technically sophisticated civilization from which he could learn; if Anderson travelled with similar expectations (and he too had requested to go on the adventure), his reaction to Chinese medical practice and its results would naturally differ from those of Staunton, Barrows and Gillan.

That the physical symptoms and even the outcome of a particular case of illness were reported so differently by different mission personnel illustrates the central role played by expectation in the creation of a medical culture. One of the major insights enabled by new social approaches to the history of medicine is that multiple medical cultures often existed within a society (or even a single region or city).[76] In the ante-bellum southern states, medicine took one shape for whites and another for slaves (and livestock). Different strengths and weaknesses, illnesses and immunities were expected of slaves – and the medical community prescribed divergent therapies for them.[77] Similarly, treatment which was considered appropriate for prostitutes (and for poor women in general) in nineteenth-century Britain would never have been used on middle- or upper-class females. Indeed, to treat these two groups of women in the same way was to risk iatrogenic disease.[78] The symptoms of slaves and prostitutes were read in ways distinct from the social norms; their illnesses, as much as their spoken words, transpired in dialect. Economics, too, created medical diversity, from the culture of self-medication, to that of spas and rest-cures.[79] The idea that the same symptoms can be, and often are, interpreted differently depending on their context is by now a truism; its connection with the communication of medical ideas and innovations is evident (if complex), particularly in the case of Chinese medicine. A profound language barrier, heightened by abstruse medical jargon on each side, exacerbated a clash between markedly different ideas of the body, health and disease which was itself mediated by unsatisfied expectations of the doctor–patient relationship. Anderson's satisfaction and Staunton's derision equally derived from the medical subcultures – subcultures defined as sharply by class and education as by time and place – to which they belonged.[80]

Medicine's mirror

> To travel through a fine country – to see pagodas, canals, and manufacturing towns, without being able to ask a single question, is extremely mortifying.... – what information could we derive respecting the arts and sciences in a country where we could not converse with the inhabitants? With what countenance will Lord Macartney return to Europe after his shameful treatment? No apology will satisfy. We go home – are asked what we have done. Our answer – we could not speak to the people.[81]
>
> Dr Dinwiddie, circa 1798

These different narratives of Chinese medicine, and particularly of cross-cultural medical practices and expectations, as they reflect the various changes (and local stability) of European medical culture, also act as mirrors of the history of elusive and tentative contact between China and the West. As the Macartney mission was being planned, medicine was coming into its own as a tool for demonstrating the superiority of European civilization. In China, medicine was following the inroads made by science.[82] Of course, medicine was also drafted into service as a gauge for the general character and state of civilization of the Chinese. Like foot-binding, and confinement of women, the medical ability and oddities of the Chinese were considered to reflect deeper truths about them. Michael Adas has argued that in fact technology and medicine were seen as particularly accurate gauges for the merit of a culture, and in general as providing corroboration of the superiority of Europeans and European civilization.[83] The reactions and writings produced by the elite (or like Barrow, aspirationally elite) members of the Macartney Embassy certainly fit this pattern. Several members of the party expressed their desire to debunk images of Chinese technology as sophisticated, while Barrow was explicit about his intentions to use his observations of Chinese science and technology 'to appreciate the Rank that this Extraordinary Empire may be considered to hold in the scale of Civilized Nations'. Indeed, Barrow was accused by a later author of deliberately fabricating tales to discredit Chinese artisans and natural philosophers.[84] He was not uniformly critical of Chinese skills and craftsmanship. Instead, he implied that it was the government of China, with its inadequate protection of property and insufficient rewards for invention, which stifled the nation's ingenuity. That national genius, however, was of a particular kind: 'The people discover no want of genius to conceive . . . and their imitative powers have always been acknowledged to be very great . . . The mind of the Chinese is quick and apprehensive, and his small delicate hands are formed for the execution of neat work.'[85] Despite Barrow's attribution of at least the potential for creative genius to Chinese artisans, he clearly did not consider them to be on a par with their British counterparts. Of their medicine (a professional, rather than labouring science) he was utterly disdainful:

> [T]he whole medical skill of the Chinese may be summed up in the words of the ingenious Doctor Gregory . . .'In the greatest, most ancient, and most civilized empire on the face of the earth, an empire that was great, populous, and highly civilized two thousand years ago, when this country was as savage as New Zealand is at present; no such good medical aid can be obtained among the people of it, as a smart boy of sixteen who had been but twelve months apprenticed to a good and well employed Edinburgh Surgeon, might reasonably be expected to afford.'[86]

Like Barrow, Staunton ended his summary of Chinese medicine and science on a note of dismissal. In the process, he revealed that the accredited observers (the 'gentlemen') of the mission defined science in narrow and specific terms:

It is a matter of doubt, whether natural history, natural philosophy, or chemistry, be, as sciences, much more improved than anatomy in China. There are several treatises, indeed, on particular subjects in each. The Chinese likewise possess a very voluminous encyclopaedia, containing many facts and observations relative to them; but from the few researches the gentlemen of the Embassy had leisure or opportunity to make...they perceived no traces of any general system or doctrine by which separate facts or observations were connected or compared, or the common properties of bodies ascertained by experiment; or where kindred arts were conduced on similar views; or rules framed, or deductions drawn from analogy, or principles laid down to constitute a science.[87]

This definition of science was exclusive and was rooted in an ideal certainly unfulfilled by European science. Nevertheless, based on this model, Staunton discriminated between China's pragmatic expertise and the nobler, disinterested sciences of Europe. Like its medicine, China's natural knowledge was reduced to the status of 'mechanical art' by this promoter of the sciences.[88] However useful, the haphazard fragments and particularities which China's experts could offer to Britain could not equal the scientific gifts they would be granted in return. Staunton and his companions scorned the idea of natural knowledge being sought first and foremost for functional goals. Thus the mission's natural philosopher, James Dinwiddie, who spent much of his time in China carefully observing and sketching Chinese technological achievements, was nonetheless annoyed by the questions put to him by his Chinese audience as he demonstrated the scientific machinery brought as gifts to the Court. When shown a burning glass, the Colao asked, ' "How can an enemy's town be set on fire by the lens? How will it act on a cloudy day?" ' According to Dinwiddie's biographer, this practical inquiry illustrated a dependence on 'ideas truly provoking to a European philosopher'.[89]

Still further indignation was provoked by the Emperor's much reported comment on the experiments and demonstrations performed before him. The Emperor 'looked at the lenses for not more than two minutes, and retired. When viewing the air-pump, &c. he said *These things are good enough for children.*'[90] Ironically, the British visitors routinely described the Chinese as children, precisely because they were unimpressed by the 'adult' gifts they were given: 'The Chinese act very much like small children: are as easily pleased and as soon tired.'[91] Dinwiddie explained away this reaction of rapidly failing attention and interest, by laying it the door of ignorance on the part of his audience, and poor planning on the part of his superiors:

> An ignorant people should always be taken by surprise. When a grand machine is shown all at once, and the principles of motion concealed, it seldom fails in its effect. But when it is shown piecemeal ...in short when a machine is built from its foundations before them...the machine [in this case, the grand planetarium] is much lowered in the estimation of the people about the palace...[92]

Dinwiddie, Barrow and Staunton were all intimately involved with the instruments brought to Court, and with the ideals of British science and civilization which those objects represented. Each had played a highly visible role in choosing and preparing the Emperor's gifts. Consequently, each had substantial social capital invested in the Emperor's reaction to his scientific presents.[93] Aeneas Anderson, on the other hand, was merely a valet, uninvolved in the mission's decision-making, and with no significant investment in its success. His account painted a markedly different picture of the cool Chinese response, and one which credited them with more discernment than usual:

> Several [of the optical, mechanical, and mathematical instruments]...when a trial was made of them before the mandarins, were found to fail in the operations and powers attributed to them; and others of them did not excite that surprise and admiration in the breasts of the Chinese philosophers, which Dr. Dinwiddie and Mr. Barrow expected, who immediately determined upon the ignorance that prevailed in China, and the gross obstinacy of the people.[94]

Anderson's account is indirectly supported by Macartney's admission that measured against the profusion of 'spheres, orreries, clocks, and musical automatons' he had seen at Jehol, 'our presents must shrink from comparison, and "hide their diminished heads"'.[95] Moreover, Dinwiddie and Barrow note that some of the instruments had been damaged during the journey.

Each of the narratives of China which emerged from the Macartney mission commented at length on the artisanal skills and technological innovations of the Chinese. Despite the obvious interest with which they made their observations, the accounts – public and private – written by the 'gentlemen' of the mission all devalue eastern technology in favour of western science. Pragmatism, in these narratives, stood for inadequacy not only in the knowledge-base of the society, but in its values; to be pragmatic was to be limited, even ridiculous. Anderson's tale, on the other hand, enthused over the 'indefatigable spirit of the Chinese people in all works that relate to public utility'.[96] To Anderson, and presumably to the unfortunate Eades, such an approach was a reasonable and appropriate response to the availability of new knowledge. Like its material technology, China's medicine was certainly marked by empiricism and pragmatism; it was also structured by a rich foundation of theory. However, this foundation was essentially unavailable to the Britons who either witnessed Chinese medical practice, or experienced it directly. Thus, they interpreted it instead in the light of their personal assumptions and expectations. The pragmatism, so limiting in the eyes of the 'gentlemen', cast its shadow onto Chinese diagnosis, and what was a theoretically grounded response to a set of sensible or legible symptoms appeared to Staunton and Gillan as a transparent attempt at obfuscation, designed to maintain the status of the physician rather than the health of the client. Anderson, with fewer objections to blunt practicality, saw a successful cure where his social superiors saw a fortunate escape from an incompetent empiric. Diagnostic competence was transmuted into arrogance – and ignorance – in the Colao's bedchamber, and Barrow and his Chinese physician both regarded themselves as the man informed and in charge of Barrow's illness. The barrier between these two geographically separated cultures of expertise was clearly constructed as much by differences in medical traditions, expectations and interpretations, as by distance, language and economic interest.

2
The Needle Transfixed: Ten Rhyne, Kaempfer and the European Gaze[1]

> I suppose my readers will be pleas'd to practice according to the Chinese mode, as well as to adorn their houses with their curious manufactures, and to use their diet of Thea.[2]
>
> Sir John Floyer, 1701

Despite his sensitivity to trends in British medicine, Sir John Floyer's 1701 prediction that Chinese medicine would be as popular in Britain as Chinese silks and tea proved over-hasty. Medical treatment was clearly a consumer good in this period, and was as much subject to the whims of its consumers as any other product in the eighteenth-century marketplace; yet in some integral way, Chinese medicine differed from tea and silk, china and lacquer-ware.[3] The exotic was less appealing in the sickroom than in the coffeehouse or at the breakfast table. In addition, information about Chinese medicine diffused through Britain far more slowly than did Chinese consumer goods. Thus, Dr Gillan of the Macartney mission shared neither Floyer's enthusiasm for the pulse nor his broad awareness of Chinese medical practice and theory; however, he and his companions of every rank did hold strong opinions and preferences about their tea. Notwithstanding an additional ninety years of western contact with China and Chinese medicine, Gillan lacked even a name for the medical technique he witnessed in Ho-Shen's chambers. Only Staunton's descriptions of the tools employed in the operation confirm its identity as acupuncture: 'The operation had been frequently performed, and many deep punctures made with gold and silver needles (which two metals only are admissible for the purpose).'[4]

Staunton's description naturally focused on the one aspect of acupuncture which was readily accessible to an observer unable to speak

Chinese. But his focus on the physical manifestations of Chinese therapeutics also reflects the materialist bias of eighteenth-century western medicine. Had they wished to do so, the British observers could have asked their interpreter to seek an explanation of the needling; they did not (or they uncharacteristically wrote off both query and reply as unworthy of comment). In describing acupuncture as intended 'to expel the vapour or spirit . . . [by] opening passages for its escape, directly though the parts affected,' Gillan and Staunton were following the western pattern of rendering Chinese medical theory into materialistic terms, despite (or perhaps because of) the nonsensical results of that rendering. This interpretive mode, combined with the increasingly science-based language of western medicine, made the Chinese explanations of acupuncture's *modus operandi* almost untranslatable.

In China and Japan, acupuncture was a complex entity based on a theoretical understanding of the body and its functioning; this model of physiology and anatomy included invisible but empirically knowable channels and points, each of which had specific curative effects.[5] These effects were considered to be produced through point- and disease-specific needling; thus the proper site for insertion of the needle depended on the nature of the illness and not solely on its location in the body. In its native context, acupuncture was composed of three unequal parts: the theories which mediated diagnosis and constrained interpretations of the empirical evidences of health and disease; the maps which directed needle-thrusts and which were based on the theoretically structured empirical evidence; and finally, the needle and the techniques by which it was to be inserted into the cuticle. Each of the three components was essential to effective acupuncture therapy. However, in the decade before its nineteenth-century resurgence, the most accurate definition of acupuncture in contemporary English medical handbooks reductively read, 'The operation of puncturing certain parts of the body with a needle, as practised in Siam, Japan and other oriental countries, for the cure of headachs, lethargies, chronic rheumatism.'[6] Most earlier definitions had simply interpreted it as an Asian medical oddity, in which needles were more or less randomly stuck into the body. Like Staunton's description of Ho-Shen's treatment, these definitions of acupuncture eliminated or ignored the theoretical and empirical traditions guiding the needle.

This kind of evidence demonstrates the proficient transmission only of the material aspect of acupuncture; the therapeutic needle, stripped of the body-map and the theories underlying Chinese acupuncture, eventually was integrated with British medical practice, at least to some degree. What happened to the other components of this medical practice and when did they drop away? How was the separation of these interlinked and interdependent elements effected? The content and rhetoric of the texts which first described acupuncture to European audiences informed this alienation of theory from practice and model from technology. That rhetoric and content were in turn reflected and reshaped by the western audience over time in ways which illustrate the wide range of possible responses to the cross-cultural transmission of medical praxis, while demonstrating the revealing persistence of certain interpretive modes – materialism, reductionism, orientalism.

The central figures in the early history of acupuncture in western Europe are Wilhelm Ten Rhyne and Engelbert Kaempfer, two medical officers of the Dutch East India Company's settlement at Nagasaki at the end of the seventeenth century.[7] They brought back to Europe the first detailed information about needling as a therapeutic practice, and made that information available, at least to the wealthy, through their published reports. These men also played central roles in the transmission of information about moxibustion, the technique of burning cones of vegetable fibres over particular points, often in conjunction with needling. Like acupuncture, moxibustion (frequently shortened to moxa) was based on the Chinese map of the body, with specific points relating to energy channels and major organs, as well as to the body's surface. Acupuncture and moxa were used in conjunction throughout Asia, but moxa was especially visible in Japan, where European observers were frequently shocked by the number of burn-scars borne by Japanese of all classes and ages. Despite the obvious similarities between acupuncture and moxa, the techniques met with very different receptions in Europe.

Trade winds: Wilhelm Ten Rhyne and the cartography of the needle

> ... as many accidents may happen from wind in the lesser, as in the greater world ...[8]
>
> Ten Rhyne, *Philosophical Transactions*, 1683

After the 1641 institution of *sakoku*, the Dutch trading post on Deshima at Nagasaki was the only European window onto Japan.[9] Given the fashionable interest in things Asian (and the profitability of Holland's exclusive Japan trade), tales of that hidden culture were much sought and swiftly devoured by British and European audiences. Wilhelm Ten Rhyne's treatise, *Dissertatio de Arthritide: Mantissa Schematica: De Acu-*

punctura: Et Orationes Tres . . ., was published in London early in 1683.[10] The voracious demand for information about Japan is evident in the fact that by June, a review of it appeared in the *Philosophical Transactions* of the Royal Society of London, increasing the availability at least of portions of his text.[11] This collection of essays began with a treatise on the gout, in which he advocated the use of moxibustion; he followed this with maps and descriptions of acupuncture, and a group of shorter essays on various unrelated topics. *Dissertatio de Arthritide* was not the first work to mention acupuncture in the West, but it was the first to explore and describe the technique in detail.[12]

The inclusion of *De Acupunctura* with a larger work on arthritis, gout and rheumatism was by no means coincidental. Ten Rhyne was writing at the cusp of the shift from humoral and hydraulic to anatomical and physiological explanations of disease. Gout, which had fitted well into humoral medicine and the hydraulic model which updated humoralism, proved far more resistant to anatomical explanation, particularly as this shift entailed also a change in the nature of medical evidence from the scholastic reliance on classical authority to the modern emphasis on empirical proof.[13] Structural explanations for gout were not immediately apparent – indeed, as late as 1768, the cause of gout was typically described in humoral or hydraulic language: 'the putrid and foul particles absorbed into the circulation, for want of a sufficient dispumation of the blood'.[14] Moreover, gout had long been uneasily balanced, in European medical theory, between the spheres of surgery and medicine: it was considered to be a systemic disease, which would normally have placed it in the province of medicine, rather than surgery. However, gout was also seen as a beneficial illness: however painful and incapacitating, an attack of the gout relieved humoral imbalances which might otherwise cause far more serious ailments. Thus, to treat an attack of gout was to risk the patient's life and thus to break the Hippocratic Oath; a physician offering such a treatment risked being stigmatized as a quack. And yet even the more sanguine of gout's victims found their suffering inconvenient, and gout cures consequently became a profitable side-line for irregular practitioners.[15]

Surgeons, eager to extend the range and raise the status of their discipline, had often claimed that as their local applications only addressed the painful symptoms of gout, patients risked nothing by accepting surgical intervention. But the conventional surgical treatments for gout – bleeding, blistering and leeching – were all recognized to have systemic as well as local effects, and medical conservatives (including physicians) claimed that they were therefore as dangerous as purges and other medical remedies, and no more effective. As the split between empirical and scholastical interpretations of disease widened, physicians and surgeons increasingly took opposite sides. By 1741, the lines were clearly visible; John Douglas stated the case for empiricism as he argued for the primacy of surgical treatments for gout:

> The gout is an inflammation... and differs from other inflammation in little more than secundum majus et minoris. And consequently belongs solely to the surgeon's province. Is it any wonder then that the Physicians have been attempting its cure in vain for above two thousand years? Their bad success was not owing so much to the disease as to themselves, because they always began where they should have ended; they always took the stick from the wrong end, i.e. they practised from theory, instead of theorizing from practice...[16]

This emphasis on empirical evidence would also have played a role in rendering acupuncture acceptable to the surgical community; although it could not be convincingly theorized in western terms, the technique could certainly be tested, and its success evaluated empirically. Since the resemblance between persisting humoral explanations of disease and those offered by Chinese medicine could only have benefited acupuncture, Ten Rhyne's choice of gout as the context within which to introduce the technique was a particularly canny one. Positions taken on either side of the debate over gout enhanced the attractiveness of acupuncture as a treatment for that disease.

Regular practitioners of either discipline bitterly resented the proliferating 'quack' remedies – and more particularly, their ethical exclusion from lucrative sale of those cures.[17] Either group would have been eager to expand into a new market if an acceptable 'regular' practice could be invented. Moreover, Europeans in Batavia (Ten Rhyne among them) had treated gout successfully with moxa, a fact which in a garbled form had already become known in Europe, greatly enhancing the reputation of moxibustion, and indirectly, Asian medicine in general, without much improving understanding of it.[18]

Building on his experiences in Asia as well as on ancient authorities, Ten Rhyne developed a theory on the workings and causes of gout; he was convinced that wind was to blame. His interest in the effect of wind on the body long pre-dated his travels in Asia – his dissertation was entitled *De Dolore intestinorum e flatu* – but he found in Japanese practice a strong line of evidence suggesting its involvement with gout.[19] His reviewer in the *Philosophical Transactions* summarized the argument of *De Arthritide* fairly accurately:

This author treating of the gout, being unsatisfyed with the notions of other physicians . . . instead of any humor which former ages have lookt on as its cause asserts flatus or wind included between the periosteum and the bones to be the genuine producer of those intolerable pains . . . and that all the method of cure ought to tend toward the dispelling those flatus. This wind he thinks is dry, cold and malignant, conveyed by the arteries to the place affected; where forcibly separating that sensible membrane the periosteum and distending it, must needs make a very sharp pain.[20]

Ten Rhyne's interpretation of the Chinese terms describing the mechanism through which acupuncture and moxibustion produced their effects shaped his theory of the origin of gout. Reciprocally, this theory of gout (and the many diseases which he postulated as its near relatives, or as springing from the same cause) coloured his portrayal of both techniques in *De Arthritide* and *De Acupunctura*. Perhaps the most decisive force shaping his report of acupuncture and moxibustion to the West, however, was his dependence on the scanty material and somewhat unreliable interpreters available to him, supplemented only by his few first-hand observations.

Ten Rhyne frequently commented on the difficulty which he experienced in the course of acquiring and translating his Japanese source material. The process of preparing the *Mantissa Schematica* was complicated and relied on a fortuitous encounter with a Chinese-speaking Japanese physician:

I gathered and translated these into Latin with the assistance of Iwanaga Zoko, a Japanese physician who knows Chinese, and with the assistance of Monttongi Sodaio, our interpreter, who speaks faltering Dutch in half words and fragmentary expressions. I solicited Zoko for this purpose when he was sent by the Governor of Nagasaki to propose medical questions . . . and to await my response. I relied on Sodaio because, although not good at explaining terms, he was more experienced in medical matters than all the other interpreters – but he was also more cunning.[21]

Cunning was a valuable attribute in a co-operative interpreter. Translating documents out of Japanese into a European language was a capital offence for his Japanese assistants, and if Ten Rhyne had been caught soliciting such translations (or smuggling them out of the country), he would have been expelled from Japan. However, cunning could not compensate for the language barrier. Chinese medical texts were (and are) difficult to understand even for native Chinese speakers, in part because of the multiple meanings of so many central terms. As he and his co-workers carried these complex terms from Chinese to Japanese to Dutch to Latin, shedding layers of meaning in every portage, Ten Rhyne complained that his interpreters' 'inexperience and limited vocabulary in Dutch' had forced him to 'omit much that was written in Chinese in the original documents'.[22]

Ten Rhyne, like many of his contemporaries, was fascinated by the discoveries of the circulatory system; the hydraulic model of the body (itself an intermediate stage between humoral and physiological explanations of the body) was still a popular one in the medical thought of the late seventeenth century.[23] Through these pre-existing lenses, it is unsurprising that the incomplete account of Chinese medical theory which was available to him appeared to Ten Rhyne as a fairly sophisticated, if also skewed, description of the circulation of the blood. The Chinese and Japanese use of pulse diagnosis merely served to confirm his translation of the unstable and multiple Asian terms into the more rigid language of western anatomy. In the same way, his interest in the effect of flatus informed his translation of the term *qi* as 'wind'.[24] These renderings of Japanese anatomy and theory initiated a set of complicated and extremely persistent misreadings and misapplications of acupuncture in the West.

In the text appended to his four diagrams, Ten Rhyne provided his reader with the laboriously translated and then copiously annotated 'original' Chinese notes which had accompanied the images. These interpretive notes show that the major error in translation arose from the Chinese term which is today translated as 'vessels' or 'channels'. It refers to the paths linking the organs and the extremities, along which *Chi* passes and in which it is created. Ten Rhyne believed that the term 'vessels' referred variously to nerves, arteries or veins.[25] He regarded this fluidity as evidence of laxity on the part of the Chinese and Japanese physicians: 'Both the Chinese and the Japanese loosely apply the term veins to arteries'.[26] This perception that Asian medical practitioners were not sufficiently rigorous was also visible in his description of what they were expected to know: 'The Chinese physicians devote all zeal to learning with precision the courses, locations, and pulses of all the arteries [here referring to the channels] with the aid of machines and figures as well as by cutting. *They pay little attention to the remaining anatomical parts.*'[27] Having chosen to interpret the term as meaning either veins or arteries, Ten Rhyne determined that a context referring to yin (his *humidum radicale*) demanded the translation 'vein'. Conversely, in the context of yang (his *calidum naturalis*) the term would necessarily be translated as 'artery'. His description of the system of channels joining the organs and the extremities through which *Chi*, whether yin or yang, would flow was otherwise accurate. Nonetheless, by referring the channels or vessels of the original back to known physical structures, Ten Rhyne inadvertently made Chinese theory – and thus the authority of the Chinese map of the body – hostage to the anatomists.

Ten Rhyne's choice of western terms which made the system observable and disprovable was not unchallenged. His reviewer in the *Philosophical Transactions* in fact returned to the term 'vessels' to summarize the system in English, although he reported Ten Rhyne's speculative translation of the term as arteries and veins. The English reader would have found that Japanese anatomy was composed almost entirely of improbably long 'vessels' (the reviewer neglected to mention that 'feet', too, was a translation of a different unit of length, ten thumb-widths):

These vessels they say are (in all) 14, whereof 12 are internal, and two external; containing the two principles of life called by them Calidum Innatum & Humidum Radicale upon which, together with the several measures of those vessels, their physical theory of man's body depends... They argue three degrees of humidum radicale contained in three distinct vessels belonging to the arms and terminating in the breast, and also in other three vessels of the legs distributed among the bowels... There are likewise (they say) three degrees of Calidum innatum contained also in three vessels belonging to the Arms, and three others of the legs; the first five feet long, the latter eight feet on either side, both terminating in some parts of the head... As to the external, that of the native heat rises from the outward, that of the radical moisture from the inward ankle: both terminate in the eyes ... They add two other external veins: both rising from the perineum end under the nostrils, the one passing before, the other behind...

Two other sorts of vessels they assign, each 12 in number mutually connected. The first they call Kee Miak, with all the windings 162 feet long, the seat of the soul: and are supposed by the Author to be the Arteries containing the Native heat whose motion is upwards; which getting the praedominium over the radicale moisture produces diseases. The other sort, called Rack Miak [*lo mo*], destitute of soul, esteemed veins... containing the radical moisture whose motion is downward, and if equal produces health.[28]

This reviewer concisely expressed his own opinion of all these measurements and windings by describing them as 'the peculiar anatomy of Japan'. Ten Rhyne, however, recounted this construction of the body in neutral tones. His personal assessment of its viability is only visible in his decision to translate ambiguous terms by specific anatomical ones. Indeed, he asserted that, 'The various movements of the blood must be learned through the precepts and rules as layed down by the Chinese . . . if the cure is to be undertaken according to their regimen.' With this goal in mind, Ten Rhyne presented his western audience with 'authentic diagrams . . . neglected and ignored through want of an interpreter' as an explanatory device.[29] The *Mantissa Schematica* presented four images, consisting of anterior and posterior views of the acu-tracts. Two were designated by Ten Rhyne as Chinese, and two as Japanese.

These images, and especially the two Japanese figures, must have seemed very odd to western eyes. Ten Rhyne acknowledged their breach of the newly emergent European visual conventions, prefacing the maps with the note that, 'a person especially skillful in the art of anatomy will belittle the lines and the precise points of insertion and will censure the awkward presentation of the short notes on the diagrams, when these should be more closely identified with the walls of the blood vessels'.[30] Like James's scandalized response to the idea of diagnostic pulse tables, the resistance to 'precision' which Ten Rhyne anticipated here stems from the conviction that the particularity of individual bodies and illnesses was an essential feature of European medicine.

The strangeness of these images, however, did not derive entirely from their informational content; they also failed to comply with the visual conventions of the anatomical atlas. The different levels of the body – points, channels, organs and skeleton – are all presented on one surface, violating the visual grammar of the contemporaneous western images, in which the corpse was 'dissected' on the page. The western plate implied naturalism and three-dimensionality, and also the exclusivity inherent to the experience of dissection. The four Asian images, on the other hand, did not explore the body beneath the skin by following the knife through it, but rather by projecting onto the body's surface the empirically or theoretically available traits of the necessarily unseen systems beneath it. They explicitly portray a symbolic body, rather than implying an actual body. On the Japanese figures (Figure 2.1), little attempt is made to individualize the body (although the very European facial features could easily have been transposed from the engraver's frontispiece portrait of Ten Rhyne himself) – unlike Western depictions of the dissected corpse, where the face and limbs were often a careful portrait.[31]

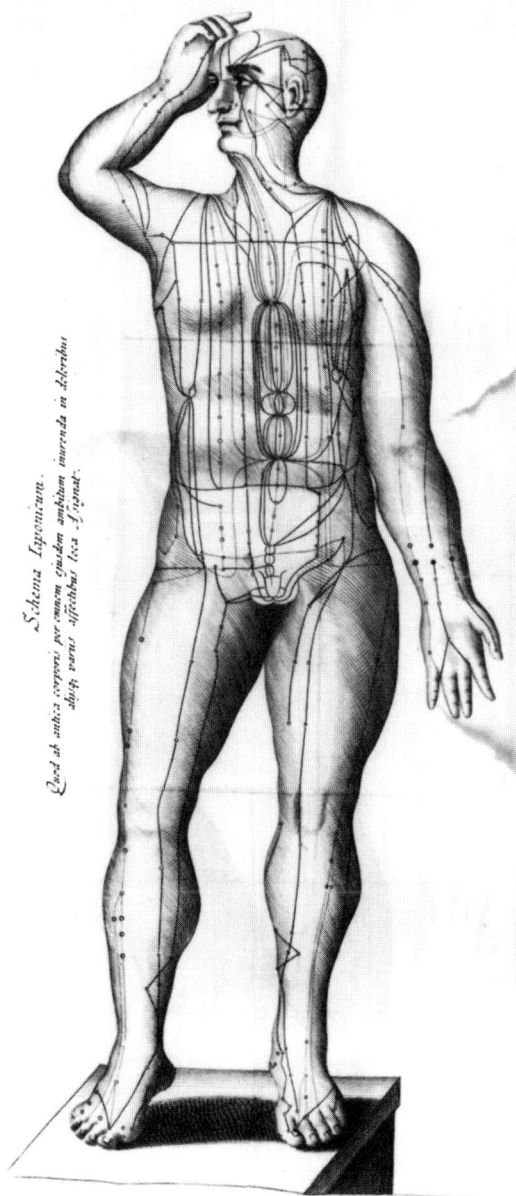

Schema Superius.

Quod ab antica corpori per unum ejusdem ambitum incurrenda in lateribus abijsq; varius difficilibus loca f̓cignat.

Figure 2.1   Ten Rhyne's 'Japanese' figure

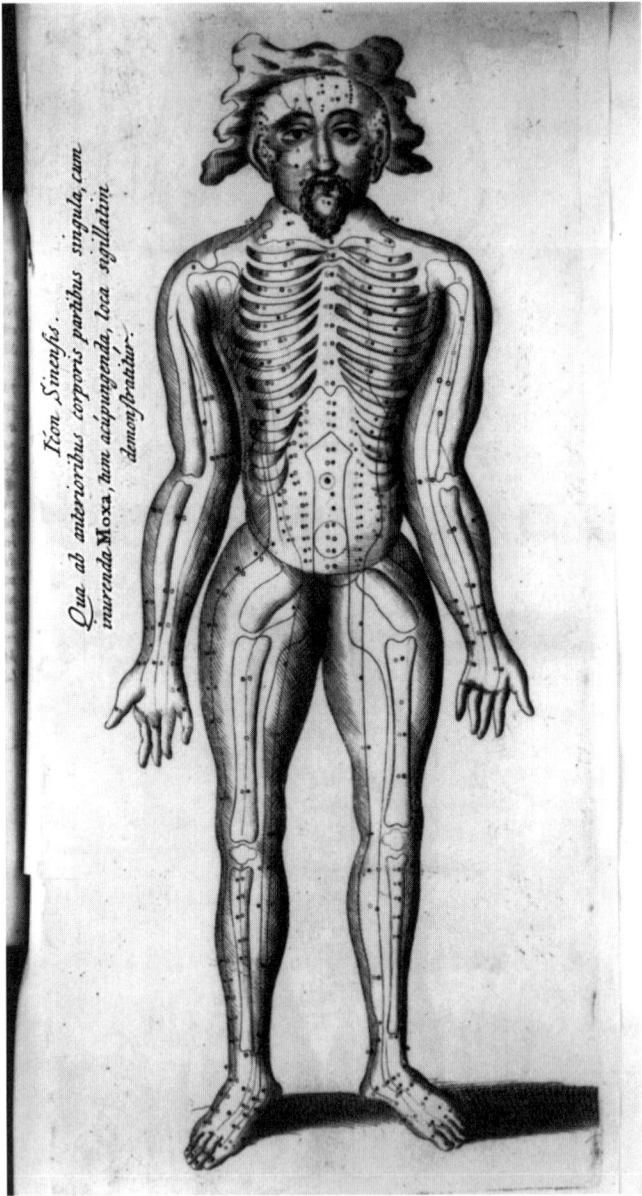

Figure 2.2 Ten Rhyne's 'Chinese' figure, front view

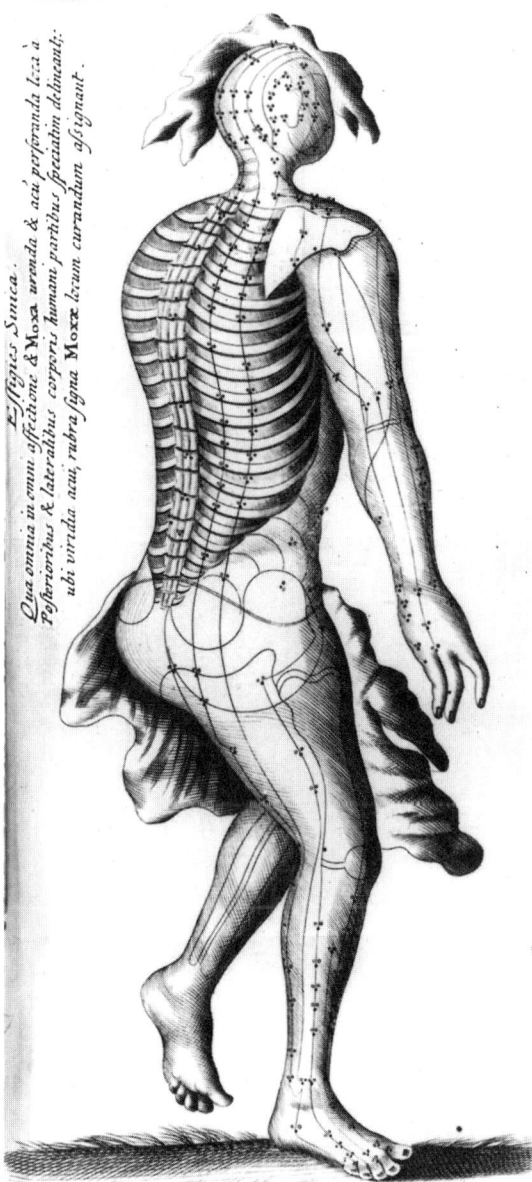

Figure 2.3   Ten Rhyne's 'Chinese' figure, back view

Figure 2.4   Wilhelm Ten Rhyne

The two 'Chinese' images (Figures 2.2 and 2.3) more closely resembled their western counterparts. The figures still convey the majority of their informational content in the form of a surface projection, rather than as a revealed inner structure. However, flaps of 'skin' have been added, and hang in rather unlikely folds around the figure, exposing the rib cage and spine, while the features of the face again seem only slightly adapted from Ten Rhyne's portrait (Figure 2.4). These innovations offer no additional information, especially since the techniques for which these diagrams were intended as guides were external applications. In fact, the originals were altered by either the engraver or Ten Rhyne himself to comply with European norms; the draped skin and facial portraiture were westernizations, designed to make the image more familiar. Similarly, the nominally Japanese pictures were shaded to create depth and perspective lacking in contemporary Japanese medical images.[32] Ten Rhyne did note that, due to 'an inconvenient location, chiefly, and perhaps also high costs', he could not include all of the individual figures he owned, and had therefore compiled four composite images.[33] His images do accurately illustrate the major channels and points, and could have guided a practitioner's needle, although they do not include information about which points were known to be efficacious in particular illnesses.

Ten Rhyne referred to the medical texts and diagrams which had fallen into his hands as 'this treasure' and was determined that it should not be lost or left idle.[34] That he took the Chinese body map seriously as a valuable aid to potential practitioners of acupuncture was evident in the analogy he used to introduce the *Mantissa Schematica*:

How does a Captain locate the harbour for his ship when he is sailing on the broad expanse of the macrocosm of the ocean? He must know how to steer a course which he plotted...on charts, to avoid by forethought sandbanks and rocks, and to calculate the probable progress of his ship...How does a practitioner discern in the very intricate circuit of the microcosm of the human body the point for burning or acupuncture...? He must understand the functioning of the heart (the regulator of our body), the position, limits, circulation and recirculations of the tiny streams of blood, and avoid injury during the operation. He must be certain of the location which each pain marks with its own sign...What other way, I ask, will a practitioner cure an ailment which yields with little difficulty to the surgical techniques of acupuncture and moxibustion?[35]

Ten Rhyne's comparison is also revealing about his opinion of the state of western medicine; in the 1680s, with the idea of the individuality and particularity of disease firmly entrenched, even the advances of anatomy could hardly reduce the sense that the body was an uncharted and vast territory. The anatomical 'atlases' which began to appear in this period were portraying the anatomized bodies of individuals, and only at the end of the eighteenth century did explicitly ideal anatomical figures appear.[36] In this introductory passage, the physician must navigate an unseen sea, guiding his patients towards health and avoiding the ever-present risk of causing harm. Only through being able to read the body as a ship's captain read his marine charts could the practitioner hope to perform this essential task. The body had been mapped by the Chinese, albeit in a strange and alien way; Ten Rhyne attributed this achievement to the extensive study of the circulation in Asia. He clearly believed that to use their technique without reference to their knowledge was as foolhardy as to sail an unknown ocean without the charts of its previous navigators. Indeed, he later compared the acupuncture needle to the magnetic compass needle.

Yet Ten Rhyne's description of the knowledge a physician needed in order to practise acupuncture and moxibustion also left room for the knowledgeable Western physician to practise independently of the Chinese maps. Such a practitioner 'must also know that it is safe to burn moxa where whirlpools of somewhat deeper blood lie concealed in fleshy areas... [and] that there is a risk involved, especially when visible articulation with its sinewy structure warns the acupuncturist to avoid a tender area, just as a ship's captain avoids a rock'.[37] In other words, a deep and specialized understanding of anatomy could replace the charted body of Asian medicine. Ten Rhyne answered his own question with the words: 'Theory furnishes laws, and experience furnishes dexterity: the best practitioner is the one who, taught and trained with both theory and experience, is a master of his art.'[38]

In discussing the Japanese maps of the body and their role in the Japanese practice of acupuncture and moxa, Ten Rhyne did not present this *particular* representation of the body as essential to the practice of acupuncture or moxibustion. Rather, he portrayed the maps as operating within a specific and limited context; drawing on ancient and long experience, they compensated for a poor knowledge of general anatomy among the physicians and surgeons of Japan and China. He (and many of his successors) also reported another role played by the maps, and one which, in western medical circles, was becoming decidedly disreputable: 'Their homes [that is, those of acupuncturists] ... have a distinctive sign:

in the entrance stands a carved statue of a human being on which are skillfully delineated the points for puncturing and burning.'[39] Ten Rhyne, in this passage, marked these three-dimensional diagrams as themselves markers, more cigar-store Indians than inalienable parts of the practice of these two techniques. They stood outside of the therapeutic chamber as 'signs' – advertisements – rather than within it as medical tools. Subsequent European authors consistently relayed this apparently insignificant detail to their audiences, suggesting that they saw encoded within it a valid comment on the worth of Japanese medical theory. Whether deliberately or inadvertently, this presentation created a gap between Japanese medical theory and practice and between the interpretations of the body which were inscribed onto the figures and the subsequent interventions upon it. Into this space, Ten Rhyne placed the corroborative systems of western medicine: ancient authority, experience and, demonstrating his awareness of medical trends in Europe, experiment. He was, nonetheless, enthusiastic in his support of acupuncture and moxibustion as practical remedies, and argued strongly for their medical value. Ten Rhyne stressed the importance of acupuncture and moxa in the Japanese medical system, emphasizing that these were their only surgical techniques, and that without them, 'their sick would be in a pitiful state, without hope of alleviation'.[40] The noteworthy health of this dependent Japanese populace was implicitly offered as proof that these were indeed 'healing methods for all pains of the body, especially pains of the external parts, which continued unresolved'.[41]

Ten Rhyne recognized several different kinds of evidence as persuasive arguments in support of acupuncture and moxibustion. His rhetoric reflected the fluidity in contemporary European medicine's intellectual and cultural foundations. He frequently called on the established authority of the Ancients – especially Hippocrates – and on the value of centuries of Japanese medical experience. However, he also employed the more modern sanction of experiment, and introduced a proto-case study as an appendix. He urged his audience to use the Japanese techniques and, anticipating doubt, combined the modern and classical forms of authority: 'Anyone who is willing to examine these matters without prejudice should test in actual practice the aforementioned canon, which has, among the ancients of the Western world, Hippocrates as its authority.'[42]

The grounds upon which he expected those doubts to be based were predominantly anatomical; having served for six months as instructor of anatomy to a group of Djakarta surgeons, Ten Rhyne was well aware of the increasing authority of anatomy in surgical circles.[43] Nonetheless, like Floyer in 1701, he expected that the combination of classical authority and the value of long experience would redeem the anatomical errors in the eyes of his audience: 'Although western anatomists may belittle these locations as inconsonant with most laws of our art, nevertheless they should not be dismissed so rashly. They have been supported by extensive experience and perfected by men of considerable acumen.'[44] At another point, he combined the modern and classical forms of medical authority, portraying the tracts and acupoints as the results of 'experiments undertaken by the very great number of superb and polished intellects of antiquity'.[45] He pointed out that Japanese ignorance of anatomy was not monolithic, reminding his readers that 'they have perhaps devoted more effort over many centuries to learning and teaching with very great care the circulation of the blood, than have European physicians, individually or as a group'.[46] However, Ten Rhyne also explicitly licensed his European audience to experiment with the body-map of acupuncture, gently criticizing both traditions of body-mapping as he did so – the Chinese for being over-elaborate, and the European for its exaggerated belief in its own accuracy:

> If anyone should not wish to make use of the elaborate work of the Chinese, let him collect through practice his own observations with which he may, with experience as his guide, make corrections and establish his own locations for burning. For the rest, the anatomist will readily overlook the fortuitous deviations of the lines and points depicted, if he studies the structure of the blood vessels: the structure is reticulate; the blood vessels composed of one substance, kiss and embrace one another, hence, when the situation of the blood vessels is determined under the knife, often there lie concealed other tiny fountains of blood which also are seats of pain...(although perhaps in a fashion other than the expert anatomist expects).[47]

On other occasions as well, Ten Rhyne used his descriptions of Japanese practice to critique its western analogues. In the course of introducing and explaining acupuncture, Ten Rhyne took several swipes at the European profession. He noted that the Japanese 'do not expound the rites of their art (to which they do not indiscriminately admit anyone) with verbal globs of honey, or ambiguous comparisons, nor obscure them with contrived and controversial nonsense, but mechanical devices clarify doctrinal analogy'.[48] The emerging success of mechanism as an explanatory system is visible in this remark; Ten Rhyne was impressed (and expected to impress his readers) by the fact that the

Chinese used technology to teach medicine: 'Thus...the masters use hydraulic machines to demonstrate the circulation of the blood to their disciples.'[49] He acknowledged the coexistence of superstition surrounding the Asian practice of medicine, but criticized those who dismissed Chinese and Japanese medicine without ever using it, remarking that 'The ill-considered eagerness of others to contradict is also unpleasant.'[50]

Like his successors, Ten Rhyne spent a striking amount of time describing the acupuncture needle itself, and discoursing on the theory that disease was caused by malevolent winds invading the body.[51] In these passages, he frequently called on western authorities, both ancient and modern, to validate the use of needles and moxa and to add weight to his theory of the winds. After an extended passage in which he described the horrifying *armamentarium chirurgicum* of the West, and outlined the neccessity, in surgery, of the manual skills of blacksmith, carpenter and tailor, Ten Rhyne finally turned to the many varieties of needle used in surgery. He drew a clear line between these needles, with their very specific functions, and the needle as it was used in acupuncture, employing the conceit of the obelisk:

> The needle, with which I here propose to deal, differs very greatly from all the previous ones. It is not a pyramid erected for the posthumous glory of a prince; it was fashioned to restore the faltering health of mankind. Neither is the needle a glorious and proud memorial; it was made to conquer the common enemy of our well-being (the corrupted and corrupting wind). It was not invented for a single and unique use, as were other needles. This needle is inserted with a blow, with a puncture and by rotation.[52]

This part of Ten Rhyne's argument comes the closest to portraying acupuncture as its technological element. He referred to 'the needle and its use – not any sort of needle, but the kind which no European shoemaker or tailor has ever seen or handled': the uniqueness here abides in the needle itself. Ten Rhyne then quoted several passages translated from Chinese texts describing the operation of acupuncture. The attributes of the acupuncture needles were set out in detail, from the materials of which they could be composed, to their length, and the shape of their handles. Ten Rhyne even included an illustration of the needle and the hammer which was used to insert it. He cited the long lists of diseases in which acupuncture had been successfully employed, but Ten Rhyne only commented on and endorsed its use for 'colic pain and other intestinal ailments produced by winds, spontaneous weakness also created by winds, swelling of the testicles, arthritis, and lastly for gonorrhea'.[53] All but the last of these illnesses were classically associated with winds and cold.

The process by which the needles (and to a slightly lesser degree the moxas) were applied to the body remained fairly stable through the transmission process. However, two important alterations in the practice of acupuncture did arise from Ten Rhyne's description of it. The first slippage occurred in Ten Rhyne's interpretation of his sources; he noted that according to Japanese practice 'acupuncture must be performed on that part of the body where the disease originates'. He clearly understood this to mean that the needles should be inserted at the site of the pain. This interpretation fitted well with what in Europe would be classified as an external, and thus surgical application, limited in its scope to direct action on a local problem. However, in Asian terms, the origin of an illness (for example, a *qi* blockage) would not necessarily be anywhere near the location where its effects was expressed in the language of pain. Thus the points which were linked to the site of organic origin might themselves be distant from either the organ or the ache.

The second change, which persisted into the 1820s in the medical dictionaries, was more immediately threatening to acupuncture's European future. This misunderstanding occurred in the ears of his western audience. Ten Rhyne included in his description of acupuncture the comment that the Japanese 'detest phlebotomy because, in their judgment, venesection emits both healthy and diseased blood, and thereby shortens life. The Chinese and Japanese frequently employ acupuncture and moxibustion in place of phlebotomy.' He then described the culturally conditioned reasoning behind this detestation, and the function of acupuncture in treating similar ailments by different means. It is clear both from the summary which follows this comment, and from the rest of the article, that Ten Rhyne knew acupuncture was not an Asian form of bloodletting. Nonetheless, this was the implication that many western readers and medical authorities drew from his work.

Ten Rhyne (unlike his successors) was convinced that some anatomical or physiological underpinning for the Chinese body-map – and therefore some physical evidence of acupuncture's modus operandi – would be found, probably in the near future. He acknowledged the difference between Chinese and western models of the circulatory system, but showed no decided preference for either, stressing instead the attributes common to both anatomies:

Although according to Western doctrine the structure of the vessels may be other than the Chinese and Japanese erect for themselves, it is nonetheless netlike...[W]hen the anatomist's knife uncovers the vessels, there are to be found lurking, the previously concealed branches of blood, the usual haunts of noxious winds.[54]

Western anatomy had not yet achieved exclusive authority, but it was already becoming the model against which other cultures' claims about the body were assessed; likewise, materialism was rapidly becoming the primary mode of medical interpretation and intervention. As the first substantial rendering of acupuncture, Ten Rhyne's portrayal shaped its future in Europe decisively. His conviction that a physical wind was released by the application of the needle, and passed out through the needle holes influenced later interpretations through the nineteenth century. The metaphors he used to convey this idea were simple and compelling – in one place, he likened the operation to pricking a hot sausage. Similarly, the direct physical description of the needle and of its use were retained through all subsequent accounts. His metaphorical description of the East Asian body-map, strongly inflected by his own doubts about its accuracy and centrality to the practice of acupuncture, was less effective and satisfying. It was both complex and ill-suited to the medical climate, and rapidly dropped out of subsequent European discussions of acupuncture.[55]

Engelbert Kaempfer: observation and acupuncture

I have been myself several times an eye-witness, that upon these three rows of holes, made according to the rules of the art, and to a reasonable depth, the colick Senki pains...ceased almost in an instant, as if they had been charmed away.[56]

Engelbert Kaempfer, 1728

Like Ten Rhyne's *De Arthritide*, Kaempfer's account of Japan, including his description of acupuncture, triggered a rapid response among the medical and social elites of Europe. He first published the *Amoentitatum exoticarum politico-physico-medicarum...* in 1712. This collection of essays was immediately cited by other writers, including medics. Interest in the subject among British intellectuals was great, as illustrated by the avidity with which Sir Hans Sloane purchased Kaempfer's entire collection and all of his writings. Sloane had the contents of the *Amoentitatum*

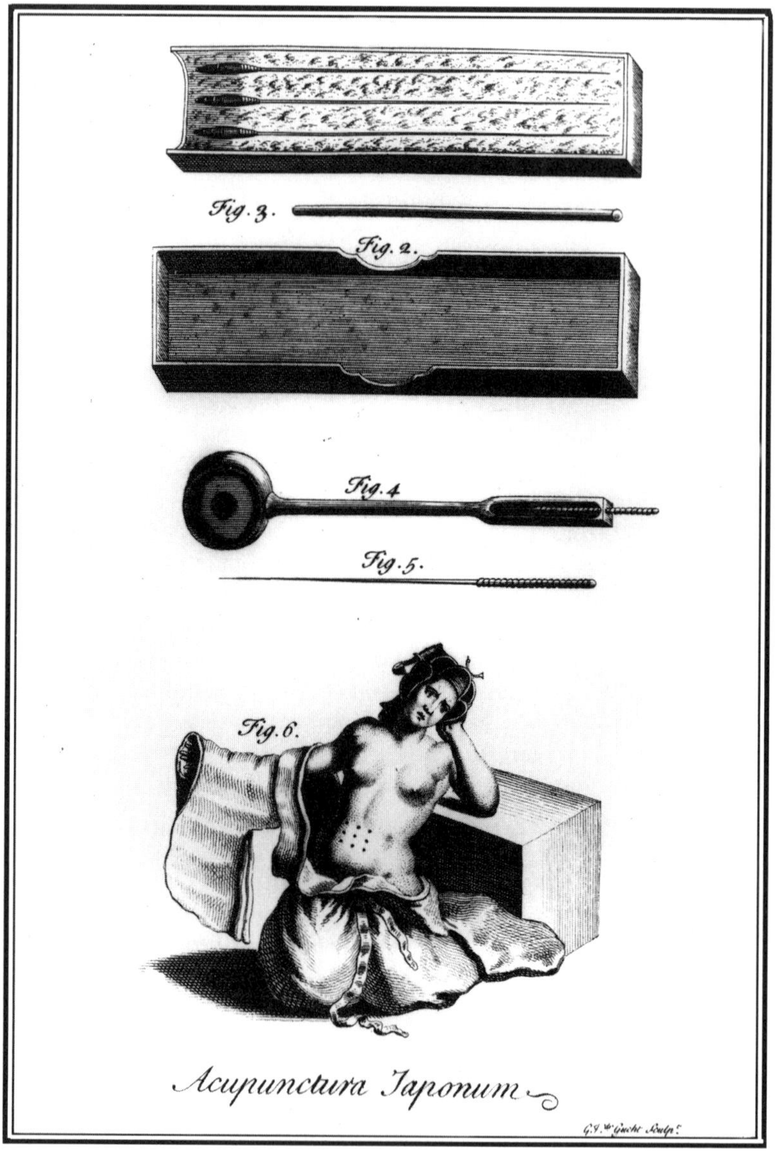

Figure 2.5   Acupuncture needles, case and patient

exoticarum translated and lavishly republished in 1728 as an appendix to Kaempfer's masterwork, the *History of Japan*. The first edition of the *History of Japan* (which Kaempfer had called 'Japan Today') was offered by subscription; among its subscribers were twenty-seven medical men, predominately the elite – members of the Royal Society, or graduates of Oxford or Cambridge.[57] Only four subscribers were surgeons, illustrating the difference in status and education (as well as income) between the two groups.

Kaempfer's account of acupuncture resembled Ten Rhyne's in terms of the information it provided about the practice of acupuncture and moxibustion. However, both tone and focus shifted sharply. Kaempfer did not attempt to justify the use of an instrument such as the needle by citing historic precedents, nor did he compare the needle to the bloodier pieces of the western surgeon's equipment. He did describe the needle itself, and the various accessories with which it was used – small hammers, cases, guiding tubes and the like (see Figure 2.5). However, he passed swiftly on to describe, in far greater detail than Ten Rhyne, exactly how to insert the needle into the patient's body. He was clear about the issue of placement and, crucially, about the very mechanical aim of puncturing:

> it [the needle] runs into the place, where the cause of the pain and distemper is supposed to be hid, where he [the practitioner] holds it, till the patient has breathed once or twice, and then drawing it out, compresses the part with his finger, by this means, as it were, to squeeze out the vapour and spirit...[58]

Kaempfer's straightforward and minute description of the actual use of the needle, unembellished by theory or explanation, was characteristic of his response of Japanese medical knowledge. He used fewer metaphors and analogies than Ten Rhyne, and made few references to ancient authority. When he did call upon classical medicine, it played an explanatory as much as an authorizing role. Thus, when Kaempfer introduced the use of moxa with a comment on the shared explanation for disease in the medical cultures of 'the Arabs, Bramines, and Chinese... [the] three chief seats of the Eastern muses', he noted a classical precedent:

> ...being ask'd their opinion about the causes of distempers, they have so frequent a recourse to winds and vapours, that they seem, in imitation of our divine *Hippocrates*... to look upon them as the

general causes of almost all diseases incident to human bodies, particularly those which are attended with pain.[59]

Here, Kaempfer's citation of Hippocrates was intended to domesticate what at first glance was a very foreign idea. When arguing for the use of the more gentle moxibustion instead of harsh caustic chemicals, he observed that the classical physicians had used dried vegetable matter to burn their patients. He did not, however, join Ten Rhyne in quoting these ancients as irrefutable authorities; instead, he presented eyewitness evidence in support of moxa's gentleness.

Kaempfer's account also makes the intersection between materialism and mistranslation even more clear. He described the way in which the Japanese understood the operation of acupuncture and moxa with remarkable accuracy, yet interpreted it – line by line – materialistically: 'They make use of two external Remedies, Fire and the Needle, both of which are thought very efficacious, to exterminate the causes of distempers, (which they call Obstructions) and to give room to the obstructing matter, as the cause of pain (which they call wind) to escape from its prison.'[60] Kaempfer was certainly better served by his translators, since the concept of clearing obstructions in the flow of some substance is a more accurate rendering of Japanese theory than the idea of opening an actual vent for some physical gas. However, he maintained the materialistic translation of the polysemic term *qi* by the word 'wind', making the more sophisticated metaphor of obstructions or energy blockages unassimilable. Instead, Kaempfer returned to the explanatory form put forward in Ten Rhyne's account, albeit elaborated to fit the anatomical standard of the day:

> As to the cause of it, and of colicks in general, the natives are of the opinion, that it is not at all a morbific matter lodged in the cavity of the guts...but that the seat of it is in the membranous substance of some other part of the abdomen...and that by stagnating there, it turns into a vapour, or rather into a very sharp sower [sic] spirit, as they express it themselves, which distends, cuts and corrodes the membranes wherein it is lodged. Upon this same theory is grounded their method of cure: whenever this spirit is let out of the narrow prison it hath been confined to, and set at liberty, that very moment, they say, the pain which it hath occasioned by distending those sensible parts wherein it lay, must cease.[61]

Unlike Ten Rhyne, Kaempfer did not believe that an anatomical basis for the Chinese and Japanese maps of the body would be found. It is worth

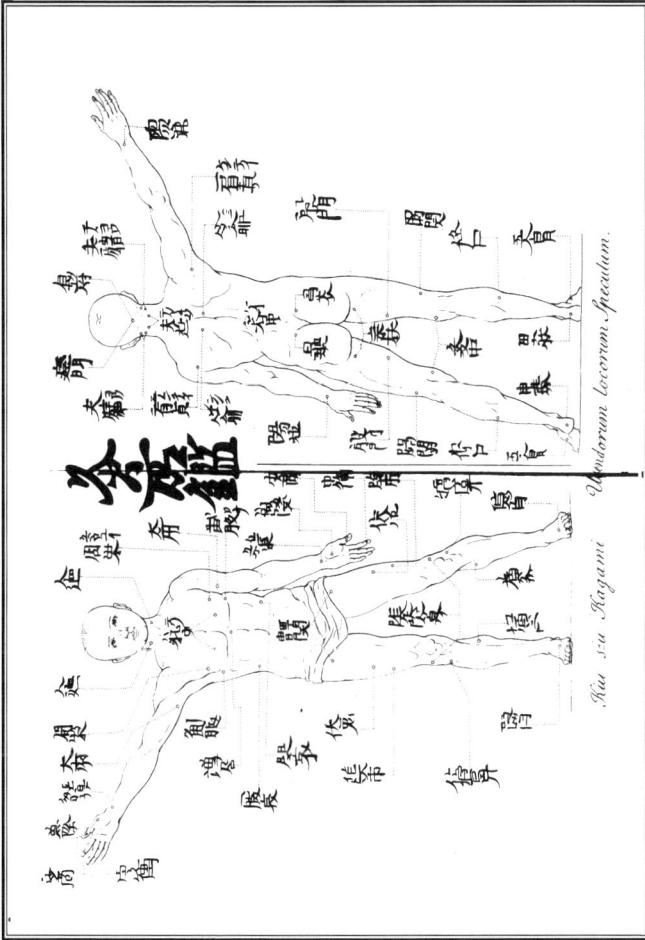

Figure 2.6   Kaempfer's acupuncture map

noting that he studied at Uppsala with Olof Rudbeck, an anatomist and a 'scientific pioneer on lymphatic circulation'. Where Ten Rhyne had studied with Franciscus Sylvius, absorbing his fascination with iatro-chemistry and fermentation (ideas which would make the idea of mor-bid gases developing in the sick body quite compelling), Kaempfer was trained to seek and expect detailed anatomical explanations, and had a relatively advanced picture of the layers of the circulatory system.[62] His sceptical response to the Japanese explanation for acupuncture was visible in his description of the technique, especially in relation to the placement of the needles: 'The precepts and rules of this pricking art are very different, with regard chiefly to the hidden vapours...the *supposed cause* of the distemper.'[63] This scepticism was far less in evidence when he wrote about moxibustion.

Another example of his scepticism – and one which derives from his reliance on anatomical understandings of the body and disease – lies in his treatment of the ideas of acu-tracts or channels, and the role of remote needling. Kaempfer explicitly noted the importance in Japanese medicine of properly placing needles or moxa, informing his readers that 'the main business lies in the choice of the part, on which either of these operations is to be performed'.[64] But Kaempfer's images, although accurately locating the acupuncture points, did not illustrate the acu-tracts on which they were placed (see Figure 2.6). In erasing this admit-tedly confusing and unorthodox surface map, the rationale underlying precise placement was also obliterated.

Kaempfer certainly knew that the Japanese hypothesized remote ori-gins for diseases:

> The Main art lies in the knowledge of the parts, which it is proper to burn in particular distempers....one would reasonably imagine [based on idea that moxa is drawing out the humours] that place to be the most proper which is nearest to the affected part, yet the operators frequently choose such others, as are not only very remote from it, but would be found, upon an Anatomical inquiry, to have scarce any communication with it, no more than by the common integuments.[65]

Kaempfer offered his audience the examples of indigestion, where moxas were placed on shoulders; cases of pleurisy, where treatment was applied to the vertebrae; and toothache, where moxas burned the adductor muscle of the thumb. In reporting this remote treatment of pain, he noted with some disdain, 'I am sensible, that the most skillful Anatomist would be at a loss to find out any particular correspondence of these remote and differing parts with one another.'[66] Where Ten Rhyne had been willing to experiment, Kaempfer was incredulous. In his dissertation, published just after his return to Europe, Kaempfer was slightly more forthcoming; he offered some information about Japanese explanations for remote treatment, but his doubts were very much in evidence:

> The place subjected to the tinder harmonises with the affected part, though there be very often no known anatomical connection... Considering the places cauterised, you would think the unexpected successes illusory... The results do not allow us to accuse them all of deception, yet sound reasoning does not permit us to testify in defence of all of them.[67]

Kaempfer simply could not accept this essential element of the practice of acupuncture and moxibustion, precisely because there was no anatomical explanation or structure underlying it. The last sentence of this passage, expressing simultaenously Kaempfer's incredulity about the claims and his trust in the claimants' veracity, was echoed with uncanny accuracy by English medics describing the puzzling efficacy of French acupuncture a century later.

Through the writings of Kaempfer and Ten Rhyne, western audiences were exposed to a progression of images and explanations of the Asian body map and the therapies based on it. Presentations of Asian medicine ranged from that of Andreas Cleyer (drawn from the work of the Jesuits in China), who included images both of the viscera and of the acu-tracts connected with them, but did not describe acupuncture; to Ten Rhyne, who described acupuncture and illustrated the channels and vessels of Chinese anatomy, but failed to grasp the mechanism of remote needling; to Kaempfer, who erased the channels, despite being aware – and telling his audience – that they played a crucial role in the Japanese understanding and practice of acupuncture. No single account was complete. More importantly, the authors who presented the practice of acupuncture in the greatest detail – the medically trained observers, Ten Rhyne and Kaempfer – were the least nuanced and most materialistic in their accounts of Chinese medical theory. This early separation of theory from practice was to have an enduring impact on acupuncture's European reception.

Strangely familiar: western responses to moxibustion

> ...for the way of curing by fire, I found twenty things to give
> me an opinion of it....[68]
>
> Sir William Temple, 1680

News of moxibustion reached Europe more than a decade before acupuncture made its appearance, largely through the experience of the Reverend Hermann Busschof, whose gout was cured by a native healer in Batavia through the use of moxa.[69] In 1677, the diplomat Sir William Temple was stricken by the gout. Having despaired of his physicians' competence to cure it, Temple was tempted by an acquaintance to try the remedy Busschof had portrayed so favourably. That he did so immediately upon reading Busschof's 'ingenious little Book' about moxa is as suggestive of his dissatisfaction with European medicine as of any interest in the medical practices of the East.[70] He himself set out the reasons prompting his experiment with tart candour:

> I pretended not to judge of the Indian Philosophy, or reasonings upon the cause of the Gout; but yet thought them as probable as those of Physicians here; and liked them so much the better, because it seems their opinion in the point is general among them, as well as their manner occurring; whereas the differences among ours are almost as many in both, as there are Physicians that reason upon the causes, or practice upon the cure of that disease.[71]

Clearly, his choice was shaped as much by the distasteful professional behaviour of European practitioners as by the attractiveness of the Asian explanation of the gout ('a malignant vapour that falls upon the joynt between the bone and the skin that covers it') and moxa (seen as a source of the heat necessary to dispel that icy vapor).[72] Temple did not merely find the squabbling of his physicians repugnant; he considered their arguments and training dangerously theoretical:

> I had past Twenty years of my life, and several accidents of danger in my health, without ANY USE OF PHYSICIANS; and from some experiments of my own, as well as much reading and thought upon that subject, had reasoned myself into an opinion, that the use of them and their methods (unless in some sudden and acute disease) was itself a very great venture, and that their greatest practicers practised least upon themselves, or their friend. I had ever quarrelled with their

studying art more than nature, and applying themselves to methods, rather than to remedies; whereas the knowledge of the last is all that nine parts in ten of the world have trusted to in all ages.[73]

Busschof's book was designed to appeal precisely to these disaffected consumers; it was written as a lively Socratic dialogue, emphasizing the practical use of moxa, its efficacy, and its relative painlessness and safety. Busschof offered the same wind-based explanation for gout as Ten Rhyne, without exploring the Chinese system underlying it. Indeed, while acknowledging the importance of correctly placing the moxas, he considered the idea of precise points, determined by the nature of the disease rather than its expression in each particular case, to be ludicrous. In fact, in his opinion this map explained why, with so effective a therapy to cure it, gout still gripped Asia.

Moxibustion successfully cured Temple's gout, and his essay describing this cure was written as a grateful letter to the man who introduced him to the burning. The essay contained enough information to allow other sufferers to experiment with moxa if they chose to do so, stating emphatically that the moxas were to be applied *in loco dolenti*. It also included a final reason adduced by Temple for his decision to use moxibustion; this last argument has profound implications for the fate of acupuncture. Temple noted that he had heard of fire used by the Egyptians, and had seen slaves scarred by the marks of the cautery iron, and that he had himself been frequently burned as a child to cure his chilblains and had seen fire used for various therapeutic reasons – putrefied wounds, ulcers, and 'casual applications of fire to the lower parts' to cure frenzies. He concluded that 'it was but a tenderness to Mankind that made it less in use amongst us, and which had introduced Corrosives and Caustics to supply the place of it, which are indeed but artificial fires'.[74] This assumption of familiarity, however misplaced, gave an enormous boost to the early use of moxibustion in the West; it was, of course, completely absent from the European and British responses to acupuncture.

Temple's essay was published in 1680, by popular demand (if his publisher is to be believed). It had circulated privately for several years beforehand, and was well-received and republished in several editions. Busschof's treatise, meanwhile, had been translated into English within a year of its original publication in 1676. Moxibustion was then pursued, with the addition of diagrams and some theory, in the texts of Andreas Cleyer, Ten Rhyne and Kaempfer. Further information about the substance of the moxas and their preparation was divulged, and in general it was commended. Kaempfer alone expressed concern that its powers were overrated by some reporters, including Busschof: 'Busho-fius...went too far, when he recommended the Moxa to his Country-men in Europe, as an infallible remedy for gout. I have reason to apprehend that many a patient in Germany found himself disappointed in his expectation.'[75] He mentioned that Andreas Cleyer had received a letter from a Dr Valentini, complaining about disappointing results from moxa. Kaempfer reasoned that while moxa worked admirably in the heat of the East, it could not be expected to work as efficiently in the chill of Europe, since the cold made muscles and membranes stiffer, and perspiration weaker. These accounts do give more detail about the Japanese and Chinese point maps, but with no more conviction than was seen in descriptions of their use in acupuncture. The material practice of moxibustion was central to the narratives, and certainly to the positive European response which greeted it.

In contrast to the enthusiasm with which the technique, if not the theory, of moxibustion was received, the initial response to acupuncture in Europe was at best uninterested and at worst critical. The crucial difference was in the existence within the western medical tradition of a precedent for moxibustion: actual cautery. This technique, in which the flesh was scorched, blistered or charred with either a flame or a red-hot iron, although not in vogue by the late seventeenth century, was at least familiar. Consumers, as well as physicians and surgeons, were aware that fire had a medicinal value, and in many cases, even possessed an explanatory hypothesis for how the healing effect of fire was pro-duced. Admirers of acupuncture, as portrayed especially by Ten Rhyne, could not call on such a resource. Of course, for many, the exotic nature of acupuncture had its own appeal. Isaac Vossius, who was fascinated by things Chinese in any case, was delighted with acupuncture:

Not less to be wondered at is that Surgery which they have cultivated in practice for so many centuries, especially in that perforation of all parts of the body which they do even of the very brain itself, trans-fixed from one side of the head to the other with a metal bodkin a cubit in length or longer. Such things have often been seen by us...either greatly mitigating or even totally removing by these means those pains to which the flesh is heir.[76]

His description probably did little to encourage the actual use of the technique in Europe. Needles, it seems, were quite terrifying enough, without being a cubit in length and piercing the brain.

It is puzzling to the modern reader to come across eighteenth-century descriptions of acupuncture and moxa as terrifying or frightful, and as so horrifying to their patients as to be unusable. That a culture which considered phlebotomy as an everyday preventive therapy and took mercury until the teeth rattled loose in the jaw should have turned squeamish at the point of a needle offers a striking illustration of the power of custom and familiarity, and conversely the fear of the novel in medicine – not just among doctors, but among their far more powerful clientele.[77] Kaempfer addressed this issue directly, acknowledging in his description of these 'two principal remedies in surgery' that, 'Their very names indeed will appear terrible and shocking to the reader, they being no less, than fire and metal.'[78] In reassuring his readers about the safety of acupuncture and moxibustion, he also critiqued the harsh expedients of western surgery:

> [I]t must be owned in justice to the Japanese, that they are far from admitting of all that cruel, and, one may say, barbarous apparatus of our European surgery. Red hot irons and that variety of cutting knives and other instruments requisite for our operations, a sight so terrible to behold ... are things which the Japanese are totally ignorant of.[79]

By contrast, Kaempfer urged, the Japanese techniques were gentle: 'Their fire is but moderate, it hath nothing to terrify the patient ... likewise the metals they make use of in their operations of surgery, are the very noblest of all ... gold and silver, of which they have needles ... which are finely polished, and exceedingly proper to perform the puncture in human bodies.'[80] Ten Rhyne's extraordinary passage recalling the saws, hammers, files, and above all the many varieties of needles employed in conventional western surgery was a less explicit treatment of the same issue.

Within the profession, reactions to Asian physic depended in part on shifts in opinion about how and where medical education should be administered, and on what constituted medical authority. Ten Rhyne's remarks about the education of Chinese and Japanese doctors were typical of the former debate. In respect to the latter, Pierre Bayle produced a remarkably lucid contemporary commentary in his review of an essay on Chinese medicine by Michael Boym:

> ... it is easy to see ... that the physicians of China are rather clever men. True, their theories and principles are not the clearest in the world, but if we had got hold of them under the reign of the

Philosophy of Aristotle, we should have admired them very much, and we should have found them at least as plausible and well based as our own. Unfortunately, they have reached us in Europe just at a time when the mechanick Principles invented, or revived, by our Modern Virtuosi have given us a distaste for the 'faculties' [of Galen] and for the *calidum naturalis* and the *humidum radicale* too, the great foundations of the medicine of the Chinese no less than that of the Peripateticks.[81]

Bayle's insightful response was unusual, and although he did not condemn the medical practices brought back from Asia on the basis of their newly obsolete language, he also did not recommend them.

Curiosity pricked: lay responses in France and Britain

We may no doubt be surprized to find the Chinese (who are so little versed in the science of anatomy, which is the most important part of physic for discovering the causes of diseases) reasoning as if they understood it. They supply what is wanting in this part by experience, and by their skill in determining by the pulse the disposition on the inward parts, in order to restore them to their natural state by proper medicines. And when all is done, no more sick persons die under their hands than do under those of the most able physicians in Europe.[82]

DuHalde, 1741

In general, European supporters and popularizers of Chinese and Japanese medicine (like William Temple and Isaac Vossius), did not engage with the Chinese theories and models of the body portrayed by Ten Rhyne, Kaempfer and Cleyer. They concentrated on describing the remedies themselves, and either dismissed the theories while praising the 'empirical' practices which those concepts had produced and explained, or failed to mention theory at all.[83] Critics of the therapies, however, often exposed and attacked the systems of which they were part. Like other information about the non-western world, these medical systems were used as evidence addressing the broader debates of the day. In Britain, for example, responses to Chinese five-element theory or the Japanese body map were coloured by the debates on the value of Ancient versus Modern learning.

William Wotton, who leaned towards the Moderns, used China as an example of the failures of Ancient learning (although he simultaneously attacked those who claimed for Chinese culture and discoveries precedence over the Greeks).[84] He drew upon the new materials coming from China to support the importance of Modern discoveries, especially in medicine, and simultaneously to discredit those opponents who claimed China as an exemplary nation, whether in medicine or in government:

> the Chinese physick is wonderfully commended by Dr. *Vossius* and Sir *William Temple*: [']*The Physicians excel in the knowledge of the pulse and of all simple medicines, and go little further*:['] Neither need they; [']*for in the first, they are so skillful, that they pretend not only to tell by it, how many hours or days sick man may last; but how many years a man in perfect seeming health may live; and by simples they pretend to relieve all disease that Nature will allow to be cure*.['] What this boasted skill is, may be seen in the little tracts of the Chinese physick published by Andrew Cleyer...[85]

Wotton proposed to give his audience a 'short specimen' of Chinese medical prose 'by which one may judge the rest'.[86] He translated into English several pages of Cleyer's translation of a Chinese text on Five Element theory. While other parts of Cleyer's text described quite pragmatic medical practices, this section was highly symbolic and allusive, creating a dissonance in western ears which Wotton reinforced by using very flowery translations of Cleyer's Latin. Unlike Floyer, who translated the same passages in his *Physician's Pulse Watch*, Wotton made no attempt to express the Chinese ideas in language compatible with western medicine.[87]

Wotton also drew his audience's attention to Cleyer's extensive selection of Chinese figures (which in fact depicted the acu-tracts and their related viscera), describing them as illustrative of 'tedious' Chinese notions of anatomy:

> The anatomical figures annexed to the tracts, which also were sent out of China, are so very whimsical, that a man would almost believe the whole to be a banter, if these theories were not agreeable to the occasional hints that may be found in the travels of the missionaries.[88]

Wotton had his own reasons for criticizing Cleyer's figures, which were less westernized than those used by Ten Rhyne. However, in summoning those images as the ultimate proof of the absurdity of Chinese medical knowledge, he indicated also the reaction he expected them to produce in other British readers. Wotton closed his discussion of China's ancient learning by acknowledging their undeniable expertise with simples, but his conclusions damned with faint praise indeed:

> [T]heir simple medicines...may, perhaps, be very admirable, and...a long experience may have taught the Chinese to apply with great success; and it is possible that they may sometimes give not unhappy guesses in ordinary cases, by feeling their patient's pulses; still this is little to physick as an art...[89]

In France, China and things Chinese were serving political as well as intellectual ends. The *philosophes* had seized China as a stick with which to beat their opponents, largely because the Confucian texts had been interpreted as a religion of reason and natural law.[90] New information about China flowed into Europe in the form of the *Lettres édifiantes et curieuses écrits des missions etrangères*; between 1702 and 1776, thirty-four volumes of these essays were published, and their contents were further disseminated in the publications of learned societies throughout Europe (e.g. the *Philosophical Transactions*). Jean-Baptiste DuHalde, who took over editing the *Lettres édifiantes et curieuses* with the ninth volume, presented China in a glowingly positive light. His personal contribution to this proto-sinology, *Description Géographique, Historique, Chronologique, Politique et Physique de l'Empire de la Chine* (published in 1735 and translated into English in 1741) was a major mid-century source of infomation about China, shaping opinions in both France and England.[91] His general description of Chinese medical theory and Chinese ideas of the body formed the basis of many later French treatments of the subject. Indicatively, DuHalde entitled this long and often insightful passage, 'System of the human body, &c. according to the ancient, but erroneous, Anatomy of the Chinese.' His introduction to the subject continued in much the same vein:

> It cannot be said that Medicine has been neglected by the Chinese, for they have a great number of ancient authors who treat of it...But as they were very little versed in natural philosophy, and not at all in anatomy, so that they scarcely knew the uses of the parts of the human body, and consequently were unacquainted with the causes of distempers, depending on a doubtful system of the structure of the human frame, it is no wonder they have not made the same progress in this science as our physicians in Europe.[92]

This 'doubtful system of the human frame' was in fact another, perhaps more intelligible, version of the material covered by Ten Rhyne, Cleyer and Kaempfer. DuHalde also included a translation (by the missionary Père Hervieu) of a Chinese tract on pulse diagnosis, explaining that the use of the pulse made Chinese medicine unique:

> They pretend, by the beating of the pulse only, to discover the cause of the disease, and in what part of the body it resides: in effect, their physicians predict pretty exactly all the symptoms of a disease, and it is chiefly this, that has rendered the Chinese physicians so famous in the world.[93]

He described physicians who felt the pulses and then told their patients, without questioning them at all, their symptoms and disease, a degree of authority which contemporary European doctors would certainly have envied. Accompanying this passage was an anecdote detailing a Chinese physician's successful cure of a missionary, which DuHalde characterized as one 'among the many instances that I could bring to put this assertion out of doubt'.

DuHalde's treatment of the Chinese system was, in general, a sympathetic one, and he made several efforts to explain Chinese practice in western terms. For example, in describing pulse diagnosis, he tried to rationalize the Chinese practice of feeling the pulse at more than one location: 'In the motion of the pulse, there are two things to be observed: the place where it is perceptible, and its duration: this has obliged the Chinese Physicians to point out the places in the body where the pulse may be examined, and the time of its beating.'[94] Similarly, rather than making a literal translation of Five Element theory as Wotton had done to such damaging effect, DuHalde merely sketched it in as 'the knowledge of the exterior bodies, which may cause alterations in the body of man'. He briefly listed the elements (earth, metals, air, fire and water) and their relations with organs and seasons, then noted '[the Chinese] reason in much the same manner as we do, concerning the agreement and disagreement of these elements with the body of man, to account for the alterations and diseases [co]incident thereto'.[95]

DuHalde was writing while the hydraulic metaphor and understanding of the body, in which was embedded an emphatic commitment to the union of mind and body, was still among the acceptable models for the body in Europe.[96] Within this model, with its focus on fluids and ferments, westernizations of Chinese theory were possible, and Ten Rhyne, Floyer and DuHalde explored similarities between the two systems by this means. However, hydraulic analogies were increasingly replaced by magnetic and electric models in the second half of the century.[97] These forces, although closely akin to *qi* as it was understood in Asia, produced conceptions of the body and disease less accommodating to the materialistic spin which Chinese theory had received in its initial western incarnation. 'Wind' certainly was no longer a viable agent in western understandings of disease and pain, yet that translation of *qi* had become firmly established, and was no longer questioned by European medical authors. In addition, the association of yin and yang with the old Galenic forces of *calidum innatum* and *humidum radicale* indissolubly linked Chinese theory to the declining system of academic physic.

Lay authors rarely included acupuncture in their descriptions of Chinese medicine, focusing instead on pulse diagnosis and herbal medicine. DuHalde never mentioned acupuncture explicitly. Indeed, even the existence of moxa and acupuncture was indicated only in a passing reference by a classical Chinese commentator included with a translated herbal. It described them as 'sharp instruments and matches, to expel outward distempers'. Given DuHalde's focus on China and his dependence on the accounts of the Jesuit missionaries, it is possible that he was either unaware of the techniques or considered them, despite the protestations of Kaempfer and Ten Rhyne to the contrary, exclusively Japanese. Wotton, however, was at least aware of Ten Rhyne's work; nonetheless, acupuncture – unlike Chinese 'anatomy' – was not included among his targets, perhaps because it had no European analogue with which even potentially to compete.

Acupuncture was not completely invisible in Europe, and surprisingly accurate interpretations of the available texts about it seem to have been made by a small minority of readers.[98] For example, the response of the anatomist Gerhard Von Sweiten in 1755 indicates a sophisticated and very open-minded reading of Ten Rhyne:

> The acupuncture of the Japanese and the cautery of various parts of the body with moxa seem to stimulate the nerves and thereby to alleviate pains and cramps in quite different parts of the body in a most wonderful way. It would be an extraordinarily useful enterprise if someone would take the trouble to note and investigate the marvellous communion which the nerves have with one another, and at what points certain nerves lie which when stimulated can calm the pain at distant sites. The physicians of Asia, who knew no anatomy, have by long practice identified such points.[99]

Von Sweiten was a physician as well as a scientist, and was European. Although few commentators on acupuncture combined knowledge of western and eastern understandings of the body with such aplomb and enthusiasm, medical men were certainly more aware of acupuncture as a specific medical (or, more usually, surgical) practice than were the laity, while continental Europeans were similarly advantaged in relation to the British.[100]

Eyeing the needle: medical interpretations of acupuncture

> C'est à ceux qui conoissent bien l'économie animale, & qui ont profondément médité sur la nature des maladies, à décider si nous devons regretter que ce moyen ne soit jamais employé parmi nous.[101]
>
> Felix Vicq D'Azyr, 1792

Temple, Wotton, Vossius and DuHalde built their interpretations of Asia, and of Asian medicine upon the reports of men either in China and Japan or recently returned from those countries. However, Ten Rhyne and Kaempfer were not replaced by informants with similar training and interest in medicine. As the stream of specifically medical observations dried up, and perhaps more importantly, as the vernacular secondary literature on the East expanded, these texts of reaction began to shape the response of the European medical community. Heister's *General System of Surgery*, first published in German in 1718, was a widely available and quite popular work. It was translated into English in 1743, and went through several English editions. In part because it was explicitly designed for the benefit of medical students and busy practitioners, and in part because of its availability, the *General System of Surgery* played an important role in the early British response to acupuncture. Heister described the technique concisely and exclusively in terms of its material operation:

> Somewhat akin to scarification is the famous operation of the Chinese and Japonese [sic], termed Acupuncturation. Those nations, rejecting scarification and phlebotomy as pernicious, have recourse to their Acupuncturation and Cauterization, or burning with the Moxa, as their most potent remedies in all disorders. The first of these operations they perform with a large gold or silver needle... which they strike into the flesh, either with their hand or the little hammer...[102]

Heister's representation of acupuncture makes it clear that thirty years of exposure to the technique had in no way lessened the horror of the needle. Scarification, the technique to which he likened acupuncture, was a form of bleeding in which a region was cupped and the resultant bruised swelling was then incised by a lancet or an array of lancets; nonetheless, piercing the flesh with a needle seemed to Heister 'desperate and severe' by comparison.[103] The half-page entry gave no rationale for the therapeutic action of the needles, nor did it describe the existence of specific, favoured insertion points. Heister excused his brevity by the fact that acupuncture had not 'been received by any of our European nations; and therefore as the process is so much to be abhorred, we shall not here give a prolix account thereof', and referred interested readers back to Ten Rhyne and Kaempfer. This type of caveat was common throughout the century, suggesting that medical interest in acupuncture remained at a theoretical rather than a practical level.[104]

Following the publication of DuHalde's detailed and popular account of China, and with the additional spark provided by occasional notes on Chinese medicine in the *Lettres édifiantes* which he edited, the French medical community began to show more interest in the subject of the needle. Consequently, Dujardin, in his 1774 *Histoire de la Chirurgie* gave a far more considerable treatment to acupuncture than had Heister. Moreover, Dujardin presented it in the context of Chinese medical theory. This did not necessarily work in acupuncture's favour – Dujardin repeated almost word for word DuHalde's unflattering introduction to Chinese 'anatomy': 'As they have no natural philosophy, almost no knowledge of the parts of the human body and their functions nor, consequently, of the causes of disease, their Medicine, entirely devoid of principle, is nothing but an unstructured jumble of systems, trial and error, and conjectures.'[105] Chinese 'surgery' came in for particular censure, since the anatomy on which it was based was 'usually the work of the imagination; thus nothing solid could ever be deduced from it'.[106] Despite these criticisms, Dujardin provided his audience with an outline of this 'imaginary' anatomy, again drawn largely from DuHalde's account and thus presented with some insight and sympathy. Dujardin himself conceded that though Chinese physiology 'may very well seem ridiculous and pitiable ... sometimes, faint glimmerings of truth pierce the fog'.[107] In particular, he appreciated the metaphor of the body as a lute, in which each of the parts sounded a note peculiar to itself and its condition.[108] Dujardin also included DuHalde's descriptions of the rest of Chinese diagnostics, noting that colour of eyes, and face, the sound of the voice, the state of the tongue, and what tastes were craved or loathed by the patient, were all used to determine the state of sufferer.

Dujardin strongly criticized the theoretical and diagnostic importance given to the pulse and to physical signs by the Chinese – and implicitly by modern European medicine as well. In place of those theories, he extolled the long experience of Chinese and Japanese physicians: 'Despite all the hypotheses which disfigure their empiricism, experience has sometimes been of use to the practitioners of China.'[109] It is worth noting that 'empiricism' had become a relatively positive term in French medicine, and no longer denoted quackery. Dujardin's positive use of it demonstrates a remarkable shift from the earlier focus in the European response (particularly among the laity) on theory as a mark of value.[110] After depicting several cases as evidence of his view that experience was the active force behind the evolution of the Chinese emphasis on pulse diagnosis, Dujardin turned from theory to practice, and from DuHalde's account to those of Cleyer – which he described as 'souvent ininintelli-gible' – and Ten Rhyne.

Dujardin culled various drug remedies from Cleyer's text, focusing on the more exotic examples, like the use of woman's milk to remedy opthalmia. He then fixed his gaze on moxibustion and acupuncture: '[T]hey have two other remedies, which they borrow from Surgery, and which they regard as specifics. All maladies which resist these, which are moxa and puncturing with needles, are considered incurable.'[111] He first discussed moxa and its centrality in Japanese therapeutics, noting that prisoners were given seasonal leave to have moxas applied, and that the Japanese population, almost to a person, bore scars from the treatment. To make that central role clear and tangible to his French audience, he explained that moxa was taken even by the healthy, 'in nearly the same manner that in Europe we have recourse to bleeding and purgation, to diminish plethora or prevent an overflow of humours'.[112] Dujardin repeated Ten Rhyne's caution that, although the treatment was not as painful as it sounded when sensibly applied – infants were able to bear it without wailing – it could be dangerous if used excessively.[113] Essentially, Dujardin offered his readers a convenient digest of Temple, Ten Rhyne and Busschof's descriptions of moxa, presented with mild appro-bation. Intriguingly, as he moved away from DuHalde's second-hand theoretical information to the rich mechanical detail of the primary sources, Dujardin's tone became more positive.

Of Dujardin's disquisition of acupuncture, the most important part was his interpretation of Chinese and Japanese theories on needle-placement. He reported the existence of 'singular figures' upon which the places where moxa and acupuncture could be performed were marked.[114] These points, he asserted, encompassed all of their science and skill in surgery. As well as including point maps carefully copied from Ten Rhyne's figures, Dujardin spent a considerable time describing them textually:

> They show the paths of the vessels, as they imagine [them]. The places which should be pricked are designated by the green dots, and those which should be burned, by red dots. The knowledge of these places was considered so important as to have since been set up as an art, which is excercised by a species of Experts, like, among us, the Bandagistes.[115]

Dujardin seems to have interpreted these maps, with their strange, surface-skimming lines, as expressing empirical (if inaccurate) knowledge of the nerves, and circulatory system.

> The points of application differ according to the type of illnesses, the character of the humours, and the nature of the underlying parts. The precepts of the art relate to the distribution of the vessels and the movement of the blood, which the Chinese and the Japanese understand better, Ten Rhyne claims, than any nation in Europe...[116]

He was incredulous in the face of Ten Rhyne's claim that the Japanese possessed a superior understanding of the circulation – how, he exclaimed, could such purely practical, observation-based knowledge be more advanced among a people who had never dissected a single cadaver than among the French?[117] By this reasoning, the anatomically informed would have little to gain by close study of the Asian body map. Instead, they could rely on their own superior knowledge of the nerves and circulation. However, unlike most European authors on acupuncture, Dujardin did not therefore simply dismiss the maps, or the principles which they represented. Instead, he argued that the Asian map, and especially the protocols which accompanied it, were the products of vast experience and, on that basis, was prepared to accept it as a guide:

> One need not believe that a slight deviation from the precise location would be an obstacle to the success of the remedy; however, several events prove that it is important not to stray from the principles... What one can certainly say [about the Chinese], is that, devoid as they are of anatomical knowledge, they could only have acquired

the principles upon which they base their application of moxa and needles from an infinite and unceasingly multiplied number of experiences.[118]

Dujardin's later discussion of the practice of acupuncture detailed the nature and shapes of the different types of needle and their modes of insertion. It also shed some light on exactly what aspect of the maps and diagrams he considered to illustrate the 'principles' – from which the would-be acupuncturist should not stray – of acupuncture. He frequently referred to the instructions which Ten Rhyne had published to accompany his *Mantissa Schematica*: that nerves should only be punctured superficially, for example; that tendons and sinews should be avoided; and that the illnesses of old and young, fat and thin people required different depths of needles. From the examples and cautions he gave, it seems that Dujardin saw the maps as charting locations throughout the body which were safe to puncture in the case of illness in that particular part. This interpretation was certainly a fair reading of Ten Rhyne's argument.[119] The needle is described as entering 'the diseased part', and 'the part in which the pain originates'.[120] Dujardin also repeated Ten Rhyne's description of the figures displayed outside the homes or shops of acupuncturists, and reiterated the claim that they served to notify potential clients of their skills, thus implicitly aligning them with advertisements.

Finally, Dujardin approached the question of acupuncture's mode of action. He thought little of the materialist version which Ten Rhyne put forward as the Japanese explanation – 'winds which slip between the periosteum and the bones; facts which he pretends are assured by observation'.[121] He, too, was aware that Ten Rhyne had been interested in flatus before he arrived in Japan, and took that into account in evaluating the 'wind' hypothesis.[122] Nonetheless, he recounted a lengthy anecdote told by Ten Rhyne of a moxibustion cure of gout. In this story, Ten Rhyne was the witness to an amazing overnight recovery. After describing the incident, Ten Rhyne exclaimed over the idea of an accord or sympathy between the parts of the body, depending on vessels unknown to European medicine. Dujardin did not place any greater reliance on this claim than he had on the claim that Japanese physicians understood the circulation better than Europeans. However, he did make Ten Rhyne's interjection the grounds of a strong criticism of contemporary French medicine:

[O]ur medicine has become too scattered and discursive; with us, the study of the parts has led us to neglect the practical science of the whole, or of that conspiracy between the parts, so well observed by Hippocrates and all true physicians: in this respect alone, the medicine of the Chinese, entirely empirical, wholly imperfect as it is, still in this respect, is worthy of attention.[123]

Dujardin concluded that acupuncture's *modus operandi* was probably humoral, and that its action was certainly weak: 'The puncture...very likely only acts by calling to the irritated part a greater effusion of humours, unless imagination, the dispenser of all physical and moral ills and well-being aids the action of this remedy.'[124] Although Dujardin was not wildly enthusiastic about acupuncture, he presented it as a legitimate therapy, based on experience and empirical success. His *Histoire* was an influential sourcebook on surgery in France; and Dujardin's treatment of acupuncture (like DuHalde's treatment of Chinese medical theory) shaped future accounts.

However, the sympathetic and serious approach to Chinese medicine and acupuncture taken by DuHalde and Dujardin was by no means indicative of a universal shift in attitudes – their mood was certainly not transmitted to Gillan, Staunton or Barrow of the Macartney Embassy. Nor was the translation of DuHalde's *Description* representative of all the information available in English. The most readily accessible English-language accounts of acupuncture were those in medical and surgical dictionaries and compendia. These sources followed Heister in assuming that acupuncture was a form of bloodletting.

Kaempfer and Ten Rhyne had made it clear acupuncture had nothing to do with bleeding – that the Japanese and Chinese were opposed to phlebotomy as unnecessarily draining good as well as tainted blood from the sick person. However, from the 1720s until the turn of the century, acupuncture was most commonly defined in English-language medical dictionaries as a form of bleeding: 'a particular way of bleeding, by making a great many small punctures with a sharp instrument, made of gold or silver'.[125] Entries always included the fact that this was a technique 'much practiced in Siam, Japan, and other Oriental nations' and often also mentioned details like the use of the needle 'even on the bellies of women with child', designed as much to shock as to inform their audiences.[126] Ten Rhyne's detailed physical description of the needle and of the acupuncturist's sphere of practice had been transmitted successfully; the nature and *modus operandi* of acupuncture, and the body maps which illustrated and accompanied them, had not.

The separation of the tool from the aims for which it was used had a profound effect: the types of analogy which filled the space between the

Asian mode of acupuncture and the needle's therapeutic use in Europe created conditions unfavourable to the adoption of acupuncture. I have found no record of the use of acupuncture in Britain during this period (though there are hints of acupuncture practice elsewhere in Europe during the eighteenth century). A simple explanation for acupuncture's failure to thrive in British medicine can be found in the pages of the many medical dictionaries and compendia of the eighteenth and early nineteenth centuries. Of twenty-four such works, six failed to mention acupuncture at all, and all but one of the remaining volumes described it as a form of phlebotomy. As an alternative to venesection, acupuncture had little to offer British doctors and surgeons. Bleeding in small amounts was considered ineffective in any illness, while bleeding with needles which terrified the patients would have seemed madness to men already competing with the blandishments of both regular and irregular competitors in a consumer-driven market. Moreover, those consumers who had heard of acupuncture were likely to have low opinions of it, at least if they had read of it in English. One of the few anglophone lay texts to describe the technique at all spoke of acupuncture as 'another way of bleeding, which may be called acupuncture, or pricking several holes in the part affected with a large needle'.[127] In portraying the remedy as it was practised in Japan, the authors were more prolix, paraphrasing Kaempfer's account at length. Their account of the practice itself was essentially neutral, but they set it in the familiar context of incompetent and ill-educated physicians, and a medicine without anatomy or physical knowledge. Unlike acupuncture, moxibustion was apparently in use in Europe; indicatively, the authors of the *Universal History* felt no need to define the term, when discussing its origins in Asia. All but three of the dictionaries I examined described moxa at some length, and with varying degrees of approbation.

Galvanizing the needle

> ... aussie peut-on dire que ce peuple, s'il n'est pas le plus éclairé, le plus savant, est le plus raisonnable, le plus doux, & le plus humain des peuples de la terre, celui par conséquent, qui mérite le plus d'être imité ... La médecine y est une pure et dangereuse charlatanerie; cependant les médecins excellent dans l'art de tâter le pouls, & prédisent assez bien de cette manière l'etat futur du malade.[128]
>
> Vicq D'Azyr, 1792

After DuHalde's general account of Chinese medicine and Dujardin's specific description of acupuncture, a more nuanced view of Chinese medical theory and practice was available in France. New models of the body and disease were also emerging. Over the course of the eighteenth century, medical practitioners increasingly incorporated the findings of the new morbid (or pathological) anatomy with the puzzle-solving techniques of classical diagnostics; to borrow Maulitz's distinction, 'general' pathology, rooted in Galenic humoralism, became first 'anatomical' and then 'physiological' pathology.[129] The newest models of bodily function retained humoral pathology's emphasis on the body as a system existing in dynamic equilibrium, but sought experimentally accessible physical substrates through which to elaborate that system.[130] Primary among these models were those based on the nervous system or nervous fluid, and after Galvani's 1791 report on animal electricity, on galvanism. Galvanism (or electricity) was swiftly, if controversially, identified as the active principle of the nerves – and both were commonly discussed as fluids, even imponderable fluids, especially in Britain. Acupuncture certainly benefited from the less materialistic interpretations of physiology fostered by this climate. Moreover, proponents of acupuncture could draw upon the doctrine of local sympathy which emerged as medicine and surgery were drawn together (largely by the increased status of the latter and by the political climate of the Revolution).[131] Essentially, 'local sympathy' was a medical interpretation of surgery's specific anatomy, filtered through the older conceptions of humoral pathology and physiology. Such a model offered at least the potential for local interventions to have systemic effects. Almost certainly also influential in the increasing visibility in the needle was the availability and popularity of mesmerism, as the similarities between the mesmeric fluid and the circulating vital energy of Chinese medicine are striking.[132] However, acupuncture was most explicitly linked, when and where the more sophisticated interpretation of Chinese medical theory was available, to galvanism and electricity; indeed acupuncture was among the first therapeutic techniques to be extensively explored in terms of its electrical effects on the body. It was this affiliation with broader medical and scientific questions and innovations which eventually enabled the practice of acupuncture to cross the Channel.

The new connection with medicine (and therefore with internal forces and bodily states) seems to have been essential to the integration of acupuncture with French practice. During this period, acupuncture and moxibustion were each defined twice in the vast reaches of the *Encyclopédie Méthodique*, both in its volumes on medicine and in those on surgery. While surgery had for a century or more been in the vanguard of medical change (and would continue to be so in Britain for several decades), French medicine was only just beginning to produce more radical offshoots. This shift was exemplified in the treatment of acupuncture in the *Encyclopédie Méthodique*. In de la Roche and Petit-Radel's *Chirurgie*, the description of acupuncture was short, adding little to Heister's coverage of ninety years earlier. Like Heister, de la Roche likened acupuncture to scarification, noted its centrality to Japanese and Chinese practice, and mentioned the precious metals of which acupuncture needles were made. The horror of these needles had seemingly decreased in the intervening years; these authors expressed shock only when discussing where those needles might be inserted: 'The nations of which we speak, although [otherwise] very industrious and sensible, perform this strange operation not only on the head, but also on the arms, legs, and other parts; they even go as far as piercing the abdomens of pregnant women.'[133] They ended their article with the dismissive comment that, 'As this operation is not practiced anywhere in Europe, we will tarry no longer here.'[134]

Moxibustion, on the other hand, was discussed at greater length, although with no new additions of information. The link between moxa and western external applications was made particularly clearly by the surgeons, to the point of suggesting that the name alone was of Asian origin, and had merely been adopted to distinguish a particular Europe practice. The remainder of the entry consisted of a lengthy quotation from Dujardin, including instructions on how to perform moxa. The authors noted the existence of point-maps, but neither exemplified nor illustrated them nor discussed their role in Japanese practice.

The volumes on medicine, edited by Félix Vicq D'Azyr, engaged with acupuncture on an entirely different level. The entry opened with a detailed comparative summary of Ten Rhyne and Kaempfer's accounts of acupuncture – critically noting those points where the two disagreed, and using those points of discordance to widen the brief of acupuncture. For instance, where Ten Rhyne reported the use of acupuncture for many illnesses and Kaempfer limited it only to the colic, *senki*, Vicq D'Azyr firmly supported the former. He reeled off lists of the conditions which the needle was used to cure in Japan, from headaches to cholera to epilepsy. He closed this list by repeating the received wisdom that, 'In all these maladies, they say, pierce the place which is the seat of the pain, or that in which the disease originates.'[135] However, the article was more than a précis of previous sources; Vicq D'Azyr intended to make his entry on acupuncture conform structurally to the model for western therapies. To that end, he organized the multifarious maladies for which acupuncture was recommended into a set of four broad Latin categories – comata, spasmi, dolores and fluxus. Having tackled the sources and applications of acupuncture, he moved on to discuss its medical roots: 'Experience taught the peoples of the Orient that... repeated and more or less profound punctures, made with needles, were a very effective aid, and that often the most acute pains were soothed immediately after they had performed this operation.'[136] He also very firmly placed both remedy and illnesses within the jurisdiction of medicine, rather than surgery.

Vicq D'Azyr described the tools of acupuncture precisely and at length, in terms of both the materials with which they were made, and their dimensions. At even greater length, he described the manner of inserting the needles. In the process, he gave a thumb-nail sketch of his interpretation of Asian medical theory, to explain why the pierced flesh was pressed subsequent to puncture. This explanation was very much in the materialist mode:

> The Chinese consider the principle of most maladies to consist of harmful vapors trapped in the suffering parts... whose release is all that is needed to heal them. This, following the system adopted by these people, is the effect acupuncture produces, by opening for these malevolent vapors favorable issues, and moxa, by attracting them to the surface of the body and consuming them there.[137]

After recommending to his readers the point-maps of Ten Rhyne, Kaempfer and DuHalde, Vicq D'Azyr turned to his own reflections on the technique. This novel section, in particular, illustrates acupuncture's changed status within French medicine. First, Vicq D'Azyr ranked acupuncture alongside western therapies: 'acupuncture is a procedure which we should rank with the irritants and stimulants... it thus can reckon with violent spasms and re-establish the sensibility in organs where that function is enfeebled'.[138] Second, while he agreed that the benefits and power of acupuncture had been exaggerated in its countries of origin ('as with famous remedies in other countries'); that its use in those countries exposed their anatomical ignorance; and even that their system of 'imaginary malevolent humours' was ill-founded, Vicq D'Azyr did not condemn acupuncture as ridiculous. Even the system, he noted, was no more absurd than many others. Instead, he handed the judgement over to 'those who know well the animal economy, and who have pondered deeply upon the nature of illnesses'. Let them decide, he declared, whether the French should regret that the technique had never been used among them.[139] In contrast to this long and innovative discussion of acupuncture and its potential, Vicq D'Azyr's treatment of moxa was desultory at best, and merely summarized the primary source material as presented by Dujardin.

By the end of the eighteenth century, it was becoming clear – at least to the French experimentalists – that acupuncture could have uses, and not just in the clinic. Vicq D'Azyr closed his essay on acupuncture with the enticing remark that 'the fact remains that these effects throw a great light on several of the most important questions in the art of healing'.[140] Vicq D'Azyr also published this essay separately, and apparently spoke on the subject to the Société de Médecine de Paris as well. Very early in the next century, a group of young French clinicians took up the idea, eager at first to explore general medical questions through acupuncture, and then to explore acupuncture itself.

L. V. J. Berlioz (father of the composer) was the first to examine acupuncture in accordance with the rules of the new clinical medicine. He ran the first known European trials of the technique in 1810 in rural Bordeaux. The results of these trials were presented in a paper given to the Société de Médecine de Paris, and were influential in interesting other experimentally inclined practitioners and clinicians – Dr Haime of Tours, Jules Cloquet, Pelletan, and others. The source of Berlioz's own interest in acupuncture is not entirely clear. His discussion of it arose in the context of a set of prize-questions about phlebotomy. Ironically, it is quite possible that the prevailing (mis)interpretation of acupuncture as an obscure method of letting blood actually led to the technique's first major success on the European stage. Berlioz himself expressed great puzzlement that a century had passed without a practical test of acupuncture's curative potential: 'The eulogia given to acupuncturation by Kaempfer and Ten Rhyne, are just and merited. We have reason to feel surprized, that although an age or more has elapsed, since this curative measure has been known in Europe, no physician has made a trial of its efficacy.'[141] Berlioz's account placed particular stress on the low risk of acupuncture, and its efficacy in nervous disorders. He then gave his colleagues two general guidelines for using acupuncture in their own practices. First, he declared acupuncture useless in any cases caused by inflammation or bloody swellings – by implication, the needle was not an apt substitute for the lancet. The second precept Berlioz proposed was that 'acupuncture, in dissipating the attacks [of illness], demonstrates that they arise from a disorder of the nervous system'.[142] In other words, if acupuncture was successful, the ailment it cured must by definition originate in a nervous disorder. This construction of acupuncture's power and actions made it useful as a diagnostic, as well as a therapeutic aid. However, Berlioz's rules also made acupuncture vulnerable to claims that it operated only by mental effects – through imagination rather than physiology. When Berlioz propounded his theories, the idea of a remedy calling upon the mind to cure the body bore little or no pejorative burden. His own laudatory words illustrate the positive light in which he viewed acupuncture's activity in these intransigent conditions: 'Simple nervous affections particularly demonstrate how much acupuncture merits the attention of physicians; because there are not many remedies which display such prompt activity and produce such marvellous effects.'[143] Berlioz offered two case studies, both as evidence for his claims and to illustrate the correct way to use acupuncture in combination with other contemporary medical therapies. Neither showed acupuncture as a miracle cure, but they demonstrated its efficacy and immediacy, and the willingness of patients to accept it, both as a medical performance and a potential cure.

Berlioz's text, although more modern in its style and content than Viq D'Azyr's article in the *Encyclopédie Méthodique*, shared with it a certain discreet relish for the exotic. One of the more striking features of Berlioz's presentation of acupuncture was its introduction. Ostensibly intended to explain how so strange a technique came into being, it gave little credit to the scientific authority or intelligence of its Asian inventors:

> The savage peoples living in the torrid and temperate zones were... in the habit of marching almost nude when they went into combat. They [therefore] experienced the necessity of imprinting on their bodies some particular signs, which... enabled them to identify themselves. The operation which they practiced to that end having been by chance done on injured parts, the resultant relief ensured its repetition in analogous circumstances. The need for signs graven on the skin having ceased with the progress of civilization, and the pricks seemingly procuring the cure only of a tiny number of maladies, the usage was lost in most nations. This remedy has been conserved only by the Chinese and the Japanese, their neighbors, where all the first institutions are sacred... It is from these people that we take the method of acupuncture: it does not belong, by any report, to the [category of] sanguinous evacuations; it can merely aid in establishing the indications for it.[144]

Acupuncture, in this tale was derived from 'savage' battle rituals and sheer chance, and was preserved by a superstitious veneration of traditions and the past. Certainly, it was no *scientia* in its native lands. This story comprised almost the entirety of Berlioz's treatment of China, Chinese medicine and the Asian origins of acupuncture. He risked reminding his readers of the East only at one other point in his narrative, in his response to criticisms of his acupuncture practice advanced by the Société de Médecine de Paris.

The Société, while awarding him an honourable mention for his essay on phlebotomy, of which his account of acupuncture formed a part, was scandalized by his liberal use of the needle, accusing him of temerity in one case.[145] Berlioz claimed that their censure weighed so heavily on him that he considered suppressing his observations entirely, but decided that they might raise questions useful to the art of medicine. Thus, he published them anyway – and defended them to the hilt. At first calling on veterinary medicine and on the harmless results of his earlier accident, he finally argued that as the Japanese had been practising in the same way for centuries, he did not consider his experiment to have been rash.

Earlier in the same book, Berlioz discussed at some length the doctrine of local sympathies, and especially their importance in the treatment of chronic illness: 'The correspondences between masses of cellular tissue must no longer be neglected in the treatment of chronic disease.'[146] Central to his version of this doctrine was the idea that the sympathetic reaction could occur at a site more or less remote from the stimulated region – and that the stimulation of the skin could activate a sympathetic reaction within particular organs.[147] He supported his interpretation with examples of sympathetic links between the shoulders and the stomach; the fat of the legs and the secreting glands of the lungs; the soles of the feet and the nasal fosses; the bronchial membrane and the intestines; and the scrotum with the gorge and all the apparatus of respiration. Berlioz even cited Erasmus Darwin's observation that a bit of ginger in the anus cured impotence as proof of his theory.[148] Obviously, such an understanding of the body was highly congenial to the adoption of acupuncture, and of its underlying body-map. However, Berlioz never mentions the existence of either specific points suitable for acupuncture, or of point-maps generally. In this reticence (or quite possibly ignorance, given his apparent initial assumption that acupuncture was a form of bleeding), Berlioz was emulated by subsequent clinicians and experimentalists.

Berlioz concluded his discussion of acupuncture with a repetition of his first rule of thumb, and further remarks on acupuncture's safety and painlessness. He called enthusiastically for further experimentation with (and on) acupuncture. In particular, he speculated that needles, combined with galvanism, in the right ventricle might re-start the heart after asphyxiation, and urged that the technique should be tested on animals (the technique was tried successfully by an experimenter some years later).[149] Berlioz closed with a direct blow at the idea that acupuncture acted by counter-irritation, and proposed his own model for its actions, as well as a way in which that model might be tested. There were, he scolded, doctors,

> who are inclined to believe that acupuncture only acts by destroying one irritation with another...to them, I repeat, it is never more successful than when it produces little or no pain. It seems, to the contrary, that this remedy acts by stimulating the nerves, or by restoring to them a principle of which they [were] deprived through the effects of the pain.... Very likely, the communication of galvanic shock produced by Volta's apparatus would increase the medical effects of acupuncture.[150]

Berlioz's insistence on linking acupuncture with galvanism foreshadowed much of the French experimental response to the treatment. In the great Paris hospital of St Louis, Pelletan and Jules Cloquet took up this thread, and began to explore the idea that galvanic fluid was in some way the explanation for acupuncture's curative effects. Their experiments, the rather inconclusive results of which were read at the Academy of Sciences, in turn prompted more interest. News of them reached Britain, and became a part of acupuncture's British form, as well as an important tool in the hands of the needle's British popularizers. Indeed, these final experimental interpretations of acupuncture, worn down to its ineradicable empirical core by a century of transmission and translation, were the foundation upon which the British practice of acupuncture was actually based, all historical lineages and claims notwithstanding. The fact that after Dujardin, none of the French authors actually included maps of the acupuncture points and channels inevitably shaped both the practice and the rhetoric of British acupuncture as it crossed the Channel and entered the public forum in the first decades of the nineteenth century.[151]

3
Sharpening the Needle: British Interpretations of Acupuncture, 1802–30

> ...how much soever our theory may exceed theirs, it will be well if their practice, upon examination, do not prove more safe and agreeable than ours, whilst they draw the main part of their medical assistance from... gentle purgatives, emollients, altera- tives, and other salubrious remedies, calculated to strengthen, rather than fatigue and weaken...[1]
>
> *Universal History*, 1759

In the eighteenth century, acupuncture was invariably described in terms of and in conjunction with its exotic Asian context, although those descriptions generally separated the technique from the theories that explained it in China and Japan. Moxibustion, on the other hand, existed in the medical literature independently of its origins. Medical dictionaries and encyclopedia, for example, often listed moxa under the techniques of actual cauterization and adustion, to which it was likened, or under the herb artemisia, from which the moxa tinders were made. By the end of the eighteenth century, moxa was familiar enough to be mentioned as a treatment option, albeit a slightly unusual one, without further explanation. Although moxa and acupuncture were first reported to the European audience in similar tones and by the same individuals, moxa had the advantage of familiarity – at least by analogy. Lacking this analogical quality, the practice of acupuncture languished, especially in Britain, despite the encouraging tones in which it too was initially reported. The eye-witness authority of Ten Rhyne and Kaemp- fer, which eventually tempted the French to experiment with acupunc- ture, evidently left the mass of British practitioners unmoved.

Unpractised, and within the British profession barely even discussed in the decades preceding its *fin-de-siècle* French rediscovery, acupuncture surfaced only in terse and inaccurate dictionary entries and often-mocking compendia articles. The first flicker of British professional interest appeared in 1802, when a surgeon in Bridgenorth – hardly a metropolitan centre – published a case study on tympany.[2] Conventionally, this disease, characterized by enormous fluid retention in the abdominal cavity, was treated by paracentesis – piercing the abdomen with lancets to release the excess fluid. Significantly, the symptoms of tympany strongly resemble those associated with the Japanese disease, *senki*, as described by Kaempfer and Ten Rhyne. With its sudden cramps, bloating, and violent evacuations, *senki* had provided the dramatic bodily context for both Ten Rhyne and Kaempfer's exegeses of acupuncture. Coley described the particulars of his case, and the successful results of the standard treatment in this instance. At the end of his study, however, he addressed a question to his colleagues:

> In the peritoneal Tympany could occasional benefit be derived by acupuncturation of the abdomen?...This operation, I believe, never was practiced much, if at all, in this country...[B]y the Chinese and Japanese it has always been an operation of practice, in a great number of disorders to which those people are liable, but particularly in a disease very analogous to tympany.[3]

Coley's question suggests that sufficient positive information about acupuncture was available to spur him to speculate on the technique, once its association with a tympany-like ailment had caught his attention. Clearly, his interest in acupuncture derived from defects in the existing therapies for tympany, and to that extent was pragmatic. However, it is also evident that Coley was aware of acupuncture essentially as a curiosity, and considered it a piece of possibly useful arcana: 'the method...is both curious and but little known'.[4]

Coley cited two sources of information about acupuncture: Heister's eighteeth-century surgical compendium, and an account of acupuncture in the popular *Universal History, From the Earliest Account of Time*. The latter was not a specialized medical publication but a popular lay resource, confirming that at least some consumers had access to information about acupuncture unmediated by medical scepticism or indifference. Indeed, the *Universal History* gave far more information about acupuncture and about Chinese and Japanese medicine generally than did Heister, and was more positive in tone. Thus, where Heister described needling as 'desperate and severe', the *Universal History* called it 'this easy and curious operation of acupuncture'.[5] In its description of Chinese medicine, the *Universal History* also revealed the degree of lay ambivalence towards the emerging medical emphasis on science. The authors reported the already familiar fact that the Chinese aversion to anatomy stemmed from their view of dissection as 'a most inhuman practice' and a means unjustified by any end, then added a barbed aside: '[They] exclaim against the anatomizing of human bodies, and *it is much to be questioned, whether the principle upon which they argue hath not saved more lives among them, than ever anatomy did among us.*'[6] Yet, despite this evident hostility to dissection, the authors concluded their critique of Chinese medical knowledge by remarking, 'we have already had occasion to hint what wretched physicians and surgeons the generality of their practitioners... are, *for want of better skill in anatomy and natural history.*'[7]

The credibility gap created by the simultaneous distaste for and fascination with the new medical materialism was shrinking by the nineteenth century. However, the popular horror of dissection remained powerful, as did the orthodox traditions of bleeding and purging which had preceded (and subsequently co-opted) anatomical and clinical observation. In their perceptions of dissection and of traditional medical practices, the British lay public were in complete agreement with their Chinese opposite numbers; thus, the reports of stasis and of resistance to anatomical study which so discredited Chinese medicine in the eyes of the emerging profession would not necessarily have harmed it in the eyes of the public. [8]

Coley's apparent aim in publishing his 'quere' was to promulgate trials of acupuncture's efficacy in Britain. Notably, although both professional and lay sourcebooks (including the *Universal History*) either implied or explicitly stated that acupuncture was a kind of bleeding, Coley made no such link, choosing instead to give no western rationale for acupuncture. Instead, he offered various pragmatic inducements to his fellow surgeons, above all emphasizing the operation's putative safety. In introducing a lengthy description of its use in Japan, he stressed the innocuousness of needling: 'As the frequency of the operation demonstrates at least its safety, it may yet perhaps be thought by the English Surgeons on some occasions to be worth imitating.'[9] For Coley, and in his opinion for surgeons in general, acupuncture was valuable and attractive as a potential substitute for an orthodox but more dangerous and complicated operation; he explicitly noted that acupuncture's simplicity should render it preferable to paracentesis.[10]

Perhaps the admirable simplicity with which acupuncture could be practised influenced his interpretation of its underlying theory; both in his description of acupuncture and in the history he quotes, it is a treatment designed 'to reach the seat of the morbific matter and... [give] it proper vent'.[11] The idea that acupuncture worked by physically allowing the body to vent its diseased contents, drawn from Ten Rhyne's original interpretation of Japanese and Chinese medical theory, had been modulated by the eighteenth-century European commentators. However, these newer interpretations of acupuncture had either not arrived in England or were unsatisfyingly controversial.[12] Coley's decision to restate the most manifest, material understanding of both term and treatment foreshadowed the British profession's response to acupuncture for the next half century. Those promoters and users who focused on this aspect of acupuncture took the term 'acu-puncture' as literally defining and delimiting the applicability and effect of the technique. Such a stance led to the development of 'acupuncture' as an alternative to lancing or otherwise physically draining the body of excess fluid or gas, as in oedema.[13] Nevertheless, Coley's article did offer more than the idea of substituting a needle for the usual lancet in paracentesis. His selection from the *Universal History* included information about the needles to be used and the importance of selecting the correct point and depth for puncture. Moreover, it offered acupuncture as an alternative to painful caustic treatments in 'topical' ailments as well:

> [T]he benefit which hath accrued from the acupuncture, in that one disease, hath encouraged others to apply it indifferently to other parts of the body... and, by a due care and precaution not to prick any nerves, tendons, or other considerable blood vessels, have cured their patients by it, without putting them to the excruciating torture which attends to that of the *Moxa*, or other caustics.[14]

The explicit attack on moxa in this passage is somewhat puzzling. When the *Universal History* was published in 1759, moxa had only recently entered the British surgical armamentarium.[15] Its prominence as a novelty, or perhaps lingering traces of its association with the Far East may explain why moxa was named as the representative caustic. Of course, its supporters presented it as less painful than the caustics and cautery to which it was compared.

Coley's article, printed over two decades before acupuncture's peak of popularity in Britain, appears as an anomaly in the chronology of British acupuncture, garnering recognition for neither its author nor its subject. The paper was never cited by Coley's successors despite their frequent recitations of acupuncture's western pedigree. Coley's article appeared in *The Medical and Physical Journal*, a prominent and highly visible publication which built its reputation on a 'preoccupation with speed and punctuality'.[16] Though not prohibitively expensive, *The Medical and Physical Journal* was not particularly cheap; in 1802, a year's subscription cost 15s 6d.[17] Its editors were also adamant in their objections to quackery, and refused to print material which they categorized as such – acupuncture clearly did not leave a bad taste even on such sensitive palates. The *Journal* encouraged reader participation, and often published reader responses. No such material followed Coley's article, indicating that acupuncture triggered neither overt hostility nor active support. How this provincial practitioner came to be interested in acupuncture, and why his article inspired no similar interest in his colleagues is unclear; perhaps his speculative tone was unconvincing. Alternatively, the fact that his query centred on the use of acupuncture for tympany, an acute ailment with a well-established (if dangerous) surgical remedy, might have limited his audience both in size and range. Whatever the reason for the article's invisibility, the fact remains that only scattered references to Chinese or Japanese medicine appeared in the English medical press between 1802 and 1822. Even such mentions as did occur were vague about the treatment they describe. For example, acupuncture *may* have been the Chinese treatment described in 1815 as 'a species of *mesmerism* or animal magnetism, as practised by certain sects of illuminati in Germany'.[18] Certainly this analogy to animal magnetism was made explicit by critics of acupuncture in later years, when both techniques were better known in Britain.

Chinese whispers: lay accounts of Chinese medicine, 1810–40

> . . . we cannot but respect the mental labours of the [Chinese] physician, though the absurdity of his doctrines, and the perversion of intellect, are rendered manifest in all his precepts.[19]
>
> Murray et al., 1836

As the Coley article suggests, the process by which acupuncture – and Chinese medicine in general – came to the attention of the British public was complex. It occurred partly through medicine and the medical press, and partly independent of it. Compendia like the *Universal History* were not the only lay sources which played significant roles;

another major source of information about acupuncture in nineteenth-century Britain was the burgeoning travel literature on China. The Macartney mission, in particular, was recognized by contemporaries as having whetted British appetites for Chinese tales. Its primary effect was: 'to draw a greater share of the public attention towards China, and to lead gradually to the study of the language, literature, institutions and manners of that vast and singular empire – a field which had hitherto been occupied almost exclusively by the French'.[20] This increased attention – and consequent consumer demand – prompted a spate of first-hand accounts from British travellers, including the members of a second official Embassy to China. Most included some discussion of Chinese science and in particular Chinese medicine, although many authors cribbed their 'observations' from the reports of Barrow and Staunton. These newer reports of Chinese medicine were rarely encouraging. Typically, they emphasized those aspects of Chinese medicine which would look most out-of-date – especially its apparent dependence on 'humoral pathology'. Even the more enthusiastic narrators tended to interpret its cures as purely empirical and largely fortuitous:

> The practice of medicine of the Chinese is entirely empirical... I had the opportunity of conversing with one of the most respectable native practitioners of Canton, and found him entirely destitute of anatomical knowledge. He was aware of the existence of such viscera as the heart, lungs, liver, spleen, and kidneys, but had no notion of their real situation, and through some strange perversity placed them all on the wrong side of the body... *Although ignorant of all rational principles of practice, he had arrived through his own experience, or that of others, at some rules of high utility; making a very clear distinction between those local diseases which can be cured by mere topical applications, and those which can only be acted upon through the medium of the constitution.* He had some vague notions of a humoral pathology, which he seemed to have perpetually in his mind whilst answering my different questions; talked of ulcers being outlets to noxious matter; and divided both his diseases and remedies into two classes, the hot and cold.[21]

Of course, in this passage, Clarke Abel, Chief Medical Officer to the Amherst diplomatic mission, told his audience somewhat less about Chinese medicine than they might have read in the eighteenth-century writings of Kaempfer, Dujardin and the Encyclopédists – but he was

British, freshly returned from China, and writing in English. The texts of this period were available, both linguistically and economically, to a far wider British audience. As in earlier encounters, many reporters expressed frustration with the inadequacy of the available translations. In this wave of communication, the culturally specific dialects of science in Britain and China received particular (if biased) attention. Abel, for example, blamed the Chinese language for the scant returns of his interviews with a Chinese physician: 'The difficulty of our intercourse, arising from the impossibility of finding adequate terms in the Chinese language for medical phrases, prevented my obtaining much accurate information respecting the details of his practice.'[22] In general, travellers tended to stress the exotic qualities of medicine, either as they observed it, or as it was described to them. Abel, in fact, drew heavily on earlier authors to supplement his more mundane observations. In quoting from DuHalde, Abel also revealed that although he was aware of moxa, specifically questioning his informant on the topic, he knew little or nothing of acupuncture: 'To act upon the imagination as well as the body, it is asserted that the part to which the moxa is to be applied is often first pricked with gold pins, and that itinerant practitioners in the north of China, fire it with much ceremony by the assistance of a convex mirror of ice.'[23] Clearly Abel had no idea that the 'pins' themselves were part of a complex therapeutic system, rather than just a bit of persuasive showmanship.

The tone in which Abel described Chinese medicine was often scoffing, matching his overall response to China. As the British presence and trade in China expanded, British attitudes towards the Chinese hardened. Responses to Chinese technology, science and medicine in particular were affected, becoming increasingly negative.[24] Moreover, some members of the British elite had already concluded that China was a known quantity, and that there was little remaining to be discovered about the Chinese. They argued that, as nothing of intellectual value had yet been exported, there was little prospect of such knowledge existing. Henry Ellis, the Third Commissioner of the Amherst mission, prefaced his published journal of that mission with the comment that, 'I much doubt the possibility of collecting any new information respecting China.'[25] As it happened, Ellis did make one unique observation relating tangentially to Chinese medicine: 'In the Koong-fu, or postures of the Tao-tse . . . their supposed influences upon disease may be traced a practice something analogous to animal magnetism.'[26] His analogy was a critical one: animal magnetism (or mesmerism) as a medical practice was being debated in Britain, with its opponents describing it as medical fraud and an incitement to indecency between medical practitioners and their female patients.[27]

Even the most sympathetic accounts of China – almost invariably found in commentaries and compendia rather than first-hand reports – were critical of contemporary Chinese medical practice. The comprehensive *Historical and Descriptive Account of China* was compiled by scholars from the Royal Society and King's College, London. It brought together information drawn from eighteenth- and nineteenth-century accounts of China, and was intended to provide a complete record of China's history, people, culture, interactions with Europe, and intellectual and commercial productions. Its authors depicted China as a pragmatic, cheerful and industrious nation, but one utterly uninterested in abstract thought. The poor progress of Chinese medicine was thus the more surprising: 'Medicine is so interesting as a science, and, as an art, so extremely useful, that we might have expected a practical people like the Chinese to have distinguished themselves by making great progress in it.'[28] Yet despite the 'voluminous works' on medicine, the imperial college sponsoring its development, and 'the early and important discoveries' about the body made in China, Chinese medicine remained inferior:

> [Its early discoveries] have been so mixed with fantastic theories and superstitious observances, that they have never been able to found upon them a system, either of sound science or successful practice. The circulation of the blood, unknown to the most learned of the ancients, and considered in Europe as perhaps the most splendid of modern discoveries, has for 2000 years been familiar to Chinese physicians. The glory, however, which they may thereby justly claim, has been greatly obscured by such irrational applications, as to have made it doubtful whether they would not have been better without possessing this knowledge.[29]

The Chinese interpretations of their lauded centuries of experience were dismissed unequivocally as unscientific – 'it is difficult to imagine that such conclusions can have been founded on any grounds either of reason or observation' – and only the desire of contemporary Chinese practitioners to learn from westerners could save them from a similarly ignoble dismissal.[30] One remark in particular is relevant to the British response to acupuncture: Murray and his colleagues noted that Chinese physicians were 'misled ... by a theory of the equilibrium of fluids invented under the dynasty of Song; and also by the application to science of the mystical cosmogony of the Yang and the Yin.' They considered these ideas, fundamental to Chinese explanations of acupuncture, to be expressions of the fact that medicine in China was 'completely under the dominion of astrology, superstition, and idolatry'. Drawing heavily on Barrow's version of the Macartney mission's encounters with 'miserable' Chinese medicine, they concluded that 'there is no nation among whom this important art is in a more inefficient state'.[31]

Despite their active support for medical missions to China, British medical periodicals showed little interest in publishing the accounts of Chinese medicine which the missionaries sent home. While missionaries might be considered authorities on the practice of acupuncture in China or Japan, their observations had no impact on British needling. Indeed, a medical text which incorporated material from Dr Robert Morrison, a prominent medical missionary in China, was clearly considered by its *Lancet* reviewer to be an oddity, 'written in the village-doctor style'.[32] Its author, Edward Sutleffe, mentions acupuncture in passing, but without reference even to the missionary account of Chinese medicine which formed the book's appendix.[33] A four-page quotation drawn from that appendix presented perhaps the most extensive extract from a missionary report of Chinese medicine to appear in Wakley's *Lancet*. Early in the passage, Morrison remarked that 'I by no means anticipate very important and useful discoveries from the Chinese in this department of human knowledge; for since the genius of Bacon threw open the gate of experimental science, the European mind has outstripped all that ever preceded it.'[34] The rest of the quote named several of the classical texts of Chinese medicine; described the Chinese view of the circulation and of yin and yang energies; and recited at some length the exotic legendary history of Chinese medicine. Like the contemporary travel narratives, the writings of medical missionaries in China placed a strong emphasis on the esoteric; their dependence on written texts over-emphasized elite academic medicine at the expense of actual medical practice. This medical orientalism also virtually obliterated contemporary Chinese practice: the use of ice mirrors was considered fabulous (and unnecessary) even by Chinese practitioners by the nineteenth century, and the ancient history of medical practices bore little relationship to their daily use.[35] Likewise, their unwavering focus on the exotic traits of Chinese practice contrasted starkly with the developing style of presentation used to introduce new techniques, including acupuncture, in the British medical press.

Acupuncture in the medical periodicals, 1810–22

Nineteenth-century commentators describing acupuncture in Asia frequently remarked on the striking fact that it did not involve any blood loss, and was not intended to do so. However, as in the eighteenth century, the medical audience was reluctant to accept their assurances. In 1822, a writer for the *London Medical and Physical Journal* included a brief comment on the subject of Chinese needling (still not identifying the practice as acupuncture) in his review of several new books on moxibustion:

> In the northern provinces of China, we are told that deep punctures are first made in the body, upon which balls of moxa are burnt. These punctures are made with needles, and the skill is to determine their number and depth. We are rather startled by the information that these *deep punctures* are not to draw blood: but this rests on the authority of the Abbé Grosier, and we fear the prop is but slender.[36]

The term 'acupuncture' was also used to describe a therapy for ophthalmic diseases; however, this 'new modification of acupuncture' did not accord well with the original. As recounted in an 1820 article, the operation instead resembled the more familiar procedures of dry cupping and scarification. Here, needling explicitly operated by 'the abstraction of blood by insertion of the needle into various parts of the body...hence the title of the process'.[37] In naming the new technique 'acupuncture', its originators were combining the literal interpretation of acupuncture as any kind of puncture with needles with the conviction that the therapy's activity depended on blood-letting.

The article in which this alternative form of acupuncture appeared was a 'Retrospective of Foreign Medical Science and Literature for the Year 1819'. Such articles were a common feature of British medical journalism throughout the nineteenth century. Typically, they consisted of information on a selected topic or innovation (as, for example, acupuncture or moxa) drawn from a range of foreign journals, summarized and annotated by an unnamed British author. In this case, the format grouped a description of acupuncture as a new sort of phlebotomy with an entirely different account of acupuncture and its uses. The second form of 'acupuncture' corresponded more closely with its Asian antecedent. The report detailed three French cases of the 'instantly successful application' of the needle for convulsions and chronic rheumatism. Paradoxically, in summarizing these case studies, the author showed some familiarity with aspects of Chinese practice, despite the contrary impression created by his earlier analogy between acupuncture and phlebotomy. The writer noted that the diseases for which acupuncture was indicated in Chinese practice ranged from coma to apoplexy, dysentery, and various 'congestions of the head and abdomen'.[38] He was even sufficiently at ease with acupuncture to extract information about its mode of action from the French original, remarking that needling could be considered 'neither an evacuant nor revulsive; since it induces no discharge, and the introduction of several needles is not more efficacious than one'.[39] In this section of the article, the reviewer also reported – without comment – certain more extreme aspects of French practice, including proposals to pierce the heart and stomach. Nonetheless, the review concluded on a positive note, emphasizing acupuncture's freedom from danger or inconvenience (despite French exuberance in its application) and its relative painlessness.

Considering the inconsistency with which acupuncture was presented even within this one article, it is unsurprising that the British profession was not convinced to adopt the technique. Of these precocious medical sources, one proposed it tentatively as an exotic substitute for the lancet; a second linked it to the already suspect French invention, animal magnetism; and the third presented it first as a form of phlebotomy and then contrarily noted that it did not act by letting blood – or in any other known way – but that it was nonetheless successful and painless. The tone of each article suggested that some low-level awareness of the technique existed within British medical community – for example, by calling the ophthalmic bleeding operation 'a new modification of acupuncture' – but all treated the term as needing definition and explanation. 'Acupuncture' was still indistinct and largely unknown to the British medical public in 1820; the technique of therapeutic needling had, however, already accumulated multiple interpretations and meanings. This multiplicity brought with it confusion and contradiction rather than flexibility. The term 'acupuncture' was stretched to cover many dissimilar uses of the needle, each of which incorporated different and often flatly contradictory instructions on how and when to apply the needle, and each of which demanded a different theoretical basis. In this respect, the British medical reception of acupuncture mirrors that afforded to other medical innovations based on or tightly linked to instruments or technologies.[40] Acupuncture's exotic origins merely enhanced this tendency by severing the technique from the explanatory resources which could have demarcated appropriate usages. Under these circumstances, the focus of acupuncture's most prominent advocates on creating a single interpretation of the term and correspondingly one unique set of rules to guide applications of the needle seems quite reasonable. As with the use of the stethoscope in diagnosis or antisepsis in surgery, British acupuncture was unified less by any consensus theory, than by the emergence of a dominant (although not exclusive) mode of practice.

'for which it is most particularly recommended': James Churchill and the singular needle

In 1822, the first English monograph on acupuncture was published. Written by a young and previously obscure surgeon named James Morss Churchill, *A Treatise on Acupuncturation* was unquestionably the most influential British work on acupuncture to appear in the nineteenth century.[41] Although Churchill opened *On Acupuncturation* with a conventional summary of the western lineage of acupuncture, his treatise rapidly diverged from its acknowledged predecessors. As his title indicated, Churchill focused on the practice of acupuncture, rather than on exotic descriptions of it or theories of its action. He highlighted the investigative and sceptical nature of his inquiry into, and the pragmatic basis of his support for 'acupuncturation':

> Under the impression . . . that there exists a desire for speculation and discovery on the one hand, regulated and qualified by a moderate and proper degree of scepticism on the other, I shall presume a medium of the two extremes, and proceed without apology or preface to my subject, trusting that the interesting facts which I have to relate will elicit such attention and investigation as will kindle a desire in some men, at least, to become acquainted with a process, which appears to rival the most successful operations for the relief of human suffering. [42]

However, before Churchill presented any 'directions for its practice' or case studies of its success, he made a determined effort to separate the practice of acupuncture, which he wished to incorporate into British medicine, from its indigestible foreign associations. The exotic quality of the treatment, its origin in the East, its luxurious accoutrements – the gold needles and ivory mallets which escaped neither Coley, nor the anonymous 1820 reviewer – all were structurally bracketed away from Churchill's argument, appearing only in a quotation describing acupuncture's Asian past. Churchill also distanced himself from the Chinese and Japanese originators of acupuncture by explicitly denying them credibility. Instead, he based his claims for acupuncture on the firmer authority of European experience:

> I should not have taken the tales which are told of the wonderful cures effected by this operation amongst the original founders of it, as sufficient authority for recommending it, nor would I admit the fables which are promulgated by these people, as evidence of its efficacy, had not this efficacy been witnessed by European spectators on its native soil, and at length experienced in our own hemisphere; and even latterly, in our own country.[43]

British eyes would, of course, see more clearly than Continental ones, let alone those of acupuncture's Asian originators. In fact, Churchill repeated this process of devaluation, even as he rhetorically wondered why acupuncture had not yet become common:

> It is a little strange, that the surprising efficacy, of which so much has been boasted by its eastern professors, and the safety at least with which acupuncturation may be performed, having been so fully demonstrated;...that it has not met with an earlier encouragement amongst us. It is probable, that the hyperbole in which it has been related, has induced the sober minds of our Northern soil, to treat these relations as the fictions of the Eastern imagination, and to reject them without examination, as fables...[44]

By this formulation, Churchill located responsibility for acupuncture's credibility gap with the therapy's apparently fabulous nature and exotic origins, and not with the British practitioners who ignored it. Moreover, Churchill pointed to the lack of critical investigation by the early supporters of acupuncture, Ten Rhyne, Kaempfer, Bidloo and Vicq D'Azyr: 'neither of them had undertaken to put its merits to the test, by actual experiment'.[45] The French, by filling this gap with experimental science, legitimated the neglected operation.[46] Thus he managed to justify previous medical resistance to even the investigative use of needling, while implying that a continuation of such resistance would be retrogressive. And yet, the ultimate validation was still to come:

> [I]t remains for the medical profession to ascertain its claims to attention by the *test of experience*, and having undergone *the ordeal of experimental inquiry*, it will, I have no doubt, so fully develope [sic]

its merit, as to obtain a conspicuous rank in medical estimation, as a valuable curative measure.[47]

It is by experience – by practice – that acupuncture was to be tested and judged; experimentation and theorization would follow naturally and in due course.

Churchill's rhetoric was of the most modern; empiricism, in its scientific sense, rang out from his introduction and its invitation to sceptical investigation. Yet this invitation was, after all, in many ways simply a longer and more detailed version of Coley's 1802 query. What made *On Acupuncturation* so much more provocative and influential? Of course, a monograph might be more visible than a lone journal article, but Coley's article was not even discussed by the readers of the journal in which it appeared. No letters responded to his query and no follow-up articles were printed. Quite the opposite reaction greeted Churchill's treatise. So how was it different? First, the call specifically to experiment was new, as was the separation of acupuncture from its Asian sources, and the subsequent bow to France. The most remarkable difference, however, lay in Churchill's attention to teaching the technique and practice of acupuncture, and in his emphasis on case studies (as opposed to testimonials or merely chained citations of authorities) as evidence.

The rhetorical efforts made by Churchill as he introduced acupuncture demonstrate the level of scepticism which he expected to greet his thesis. His response to the sceptics dwelt upon practice to the entire exclusion of theoretical concerns or appeals. His description of how to use the needles first depicted acupuncture's virtues as an alternative to more complicated surgical procedures: 'The method of performing the operation is simple and easy, requiring neither practice to give dexterity, nor adroitness that it may be done with propriety.'[48] Moreover – and this would certainly have been an attractive feature in the highly competitive medical environment of the day – Churchill observed that the use of this therapy could be limited to surgeons: 'Anatomical knowledge of the human body is, however, necessary; as an imprudent application of it, by an operator ignorant of the structure of the part into which he introduces his needle, might be productive of bad consequences.'[49] At this historical moment, of course, anatomical knowledge was still more or less the specialized fruit of a surgical education, and Churchill's ostensible aside – 'and no other ought to perform this or any operation' – was actually a strong assertion of surgical priority.[50]

Even in his use of the familiar quotation from the *Universal History*, Churchill concentrated on practice. The extract was provided only as a description of the *original* methods of applying acupuncture. It was (again) a text used to prod the curious to action, but Churchill, unlike Coley, did not offer it heuristically, as 'worth imitating', much less 'to conclude the subject'.[51] Although Asia, as the sole source of information on acupuncture's historical use, was as yet inextricable from discussions of its applicability, Asian methods of acupuncture were no longer offered as normative. Churchill's deployment of the passage implied that the outlandish attractions of acupuncture's history could and should be separated from the serious business of promulgation and popularization. It might grab the attention of a reader, but in itself was unconvincing. In fact, Churchill referred to the quotation minimally; its account of the remarkable cure of *senki* he credited only to the extent of noting that experiments with acupuncture on abdominal and visceral complaints might be informative. Churchill's only other nod to eastern expertise – and a very unusual one – related to Indian practice:

The Indians ... puncture the head in all cases of Cephalagia, in comatose affections, Ophthalmia, & c. They puncture the chest, back and abdomen, not only to relieve pain of those parts, but as a cure for Dysentery, Anorexia, Hysteria, Cholera Morbus, Iliac Pasion, & c. *Local diseases of the muscular and fibrous structures of the body also afford them occasions for its performance; and it is for diseases of this class only, that I have hitherto practised it, and for which I would expressly recommend it.* [52]

Indian practice, then, included the 'correct' use of acupuncture as therapy for rheumatism and gout. That an Indian, rather than a Chinese, precedent was cited may reflect the lingering British respect for classical Hindu culture which followed the rediscovery of ancient India's scientific and medical texts.[53] However, in Churchill's formulation, this efficacious 'native' use of the needle was presented as a merely fortuitous offshoot of the discarded practice of treating with acupuncture what were in the West regarded as systemic – and hence quintessentially medical – illnesses. Certainly, Indian practice could not be seen as truly authoritative if it was only accidentally appropriate.

For Churchill, the type of evidence which could convince his colleagues of acupuncture's efficacy was easily defined; the needle's British foundation was to be based in '[t]he success of my own subsequent practice ... [and] the opinion and experience of some physicians of eminence, accompanied by a relation of some cases of my own'.[54] Clinical success and rigorous (at least by the medical standards of the time) experimental investigation: by his own estimation, these were Churchill's most powerful tools in the promotion of a new and inescapably foreign and exotic surgical therapy. He was anxious to present the strongest possible case, to the point that he described the evidence he discarded in as much detail as that which he would eventually use. He argued that:

> Neither sufficient time has elapsed, nor a proper selection of cases been made since this operation has been known to me, to have afforded me, either a large number of experiments, or a great variety of diseases on which to try the effects of it: . . . at present *I should not be doing justice to my subject, to form conclusions on such imperfect evidence; I shall therefore confine myself, merely to the description of the good effects, which I have witnessed in diseases of a rheumatic character, and in those injuries of the fibrous structures of the body,* which are often observed to arise, (particularly in labouring persons) from violent excertion [sic]. [55]

By implication, this statement also buttressed Churchill's strategic limitation of acupuncture's applicability. In selecting rheumatic and muscular afflictions, he excluded the practice of literal 'acu-puncture' – the needle acting simply as a more delicate lancet – despite the early appearance of this usage in the medical literature.[56]

Churchill spent a substantial portion of his energies on framing acupuncture, and especially on describing its *European* roots and presence. For example, he evoked acupuncture's long, if unacknowledged, history in the European medical canon, translating portions of Berlioz's *Mémoire sur les Maladies Chroniques, les évacuations sanguines et l'acupuncture*: 'The eulogia given to acupuncturation by Koempfer and Ten-rhyne, are just and merited. We have reason to feel surprized, that although an age or more has elapsed, since this curative measure has been known in Europe, no physician has made a trial of its efficacy.'[57] Churchill quoted at length Berlioz's comments on the safety and simplicity of acupuncture, and its efficacy in nervous disorders and nervous symptoms. He then used the words of another French proponent of acupuncture, Dr Haime of Tours, to restate the difference between acupuncture and bleeding and reiterate his warning against the over-use of acupuncture, citing Chinese and Japanese practice as a bad example.

Churchill also included an extract from Haime describing French scientific investigations of acupuncture. At one point, Haime discussed the role of Berlioz's experiments in stimulating his interest in acupuncture; Churchill used this example to display first the conviction that these experiments produced in one reputable (if French) doctor, then the beneficial results that followed for the patient. Only after this presentation of his French sources and a final admiring description of their detailed experiments on acupuncture's electrical properties, did Churchill present his own cases – for Churchill, as for Gillan, experiment was clearly the hard currency of authority. With strategically understated confidence, he declared: 'I doubt not but I shall make it appear that the beneficial effects of the remedy employed, are sufficiently flattering to deserve the esteem I hold it in, and to justify me in bringing the subject into general notice.'[58]

The cases which followed offered a detailed and instructive view of Churchill's technique in using acupuncture. For the purposes of the present discussion, it is sufficient to note that the operations he performed could easily have been repeated from the information he disclosed. He did not mention either acupuncture points or acupuncture channels; instead, he provided very detailed information on the placement of his needles at the seat of pain. In the cases of acute muscular afflictions, cramps and spasms which he drew from his own practice, Churchill portrayed the effects of acupuncture as producing instantaneous and enduring relief. Churchill also offered a letter from another surgeon, Edward Jukes. In this letter, Jukes reported his own successful use of acupuncture on yet a third surgeon, Mr Scott, whom both Jukes and Churchill credited with introducing the treatment to England. As well as illustrating the kind of network which linked acupuncture users, this report included Jukes's fascinating description of the sensations produced by the acupuncture: 'I should have stated that the sensation, described as resembling somewhat an electrical effect, was experienced from two of the needles only; the first and the last of those which were introduced.'[59] Despite making this explicit link to electricity – one of the most popular (in both senses of the word) explanatory systems then available – Jukes refused to speculate on acupuncture's mode of action. In this respect, his letter exemplified what would become a trend among British users of acupuncture. However, unlike many of his colleagues, Jukes frankly admitted the reasons for his refusal:

I send you the history of this case without any comment upon the mysterious nature of this extraordinary operation; yet I am convinced there is something more in it than has been hitherto explained. I have, it is true, some notions (not however fixed) as to its nature; but *I would not at present venture to detail them, lest the embers of animal magnetism might be rekindled in the discussion, and the operation from being associated with an exploded theory, sink into an undeserved and premature oblivion,* from preconceived prejudice. [60]

Churchill, too, explicitly resisted forming a hypothesis about how acupuncture produced its cures. He concluded *On Acupuncturation* with words similar to, if less frank than Jukes's: 'I have not attempted an hypothesis of the operation. I have by no means made up my mind as to the nature of its action, and rather than venture into speculative reasoning, which would be received as doubtful by some, and visionary by others, I prefer preserving a profound silence.'[61] He noted that some of his readers would doubtless be surprised at his reticence – medicine in the Regency contained few shrinking violets. But Churchill's stance indicates a transition in British medicine. Emphasizing both experiment and experience, Churchill's rhetoric – like the decade in which he wrote his monograph – was poised equidistant between medicine conceived as an individualistic art informed by long experience, and as an increasingly uniform science in which experience was supplemental rather than central.[62]

Interpretive responses to *A Treatise On Acupuncturation*

Churchill's *Treatise On Acupuncturation*, the first full-length presentation of acupuncture to the British profession, already contained the seeds of the two major complexities surrounding the practice of acupuncture in Britain. First, acupuncture proved frustratingly intractable to orthodox medical theory and was already being perceived and presented as such. Churchill's extended title gave away his strategy for dealing with this intractability, and offered some part of an excuse for it: *Being a Description of a Surgical Operation Originally Peculiar to the Japanese and Chinese, and by them denominated Zin-King, Now Introduced into European Practice, with Directions for its Performance and Cases Illustrating its Success.* This title temptingly displayed acupuncture's alluring – but professionally undesirable – exoticism; in a market still driven by consumers, a link to the latest fashions could not lightly be ignored by promoters of a new therapy. However, the technique's exotic origins were also firmly relegated to the past. Acupuncture's prospects and status were linked instead to its use in Europe. Finally, the coda – 'with directions for its performance and cases illustrating its success' – implied that acupuncture could legitimately be taught and performed empirically, given the cases available to illustrate its success. Jukes' hints about animal magnetism indicate an additional reason for Churchill's silence on the subject: it would have been difficult to promulgate a theory tarred with the brushes of quackery and revolutionary radicalism.[63] In the absence of theory, Churchill structured his attempt to popularize the technique around creating a standard style of acupuncture practice.[64] His 'directions' were authorized by the cases in which his particular mode of practice had succeeded, while alternative modes of use were criticized either explicitly or implicitly. Tellingly, Churchill often linked opposing western interpretations of 'acupuncture' with the disreputably foreign origins of therapeutic needling. For example, in his remarks on acupuncture in India, Churchill specifically mentioned its use in abdominal, cephalagic, and ophthalmic illnesses as illustrative of the erroneous, even superstitious, Indian mode of needling. Examples from each of these disease-categories had been proposed as appropriate occasions for acupuncture by other western interpreters of the technique.[65]

Churchill's emphasis on technique hints at a second major set of issues surrounding early British acupuncture. It raises questions about what makes a medical therapy (or for that matter any other form of intimate knowledge) satisfying, assimilable, believable – and repeatable. The maps of acupuncture channels and points, and indeed any information about that aspect of needling disappeared as acupuncture crossed the Channel. The mapped body underlying Asian acupuncture had played no visible role in the French experimental exploration of the therapy, and these were the spark for British interest in the early nineteenth century. Yet without those maps, and the conjunction of theory and experience – the understanding of the body itself – which they represented, how was the technique of acupuncture to be systematized and transmitted in an effective form? In general, how important is theory in creating acceptance and naturalizing innovation, whether foreign or domestic? Conversely, how satisfying and convincing can even a successful practice be in the absence of theory?

To the extent that it was ever a mainstream technique, acupuncture was popular because it was a successful therapy.[66] Obviously, a therapy flourishes not merely because or in spite of its clinical efficacy (indeed, the new clinical medicine of the nineteenth century had a rather dubious relationship with efficacy), but also because it meets more broadly pragmatic criteria. It must be simple, easily taught and learned, offer its users – both professionals and patients – benefits not otherwise available, and especially in the pre-anaesthesia days of the early nineteenth century, be as painless as possible. Regency surgeons in particular had much to gain from a painless technique; the brutality of surgical operations was a major factor in surgery's continued low status, and in the ambiguous social standing of its practitioners.[67] But an innovative therapy must also be intellectually satisfying; ideally, both practitioners and patients must benefit from and believe in the treatment, however they interpret it. Acupuncture, as a foreign technique, had to become assimilated, to appear less absurd to the eyes of its audience. The therapeutic use of the needle had to be made intelligible in some way. Thus, Churchill presented it as readily learned and practised, as effective in intractable cases, and as exclusively surgical. He tried to create a comprehensible acupuncture by defining it as a specific, available practice, thriving, if unfortunately little known. Subsequent to the publication of *On Acupuncturation*, a string of successful case studies appeared in the medical press reporting acupuncture cures, often where the needle was a 'last resort'. In following British acupuncture from this initial, promising series of successes to its eventual slump into clinical mediocrity, it will be essential to examine the consistency and results of its technical practice as well as its fit with native medical productions and understandings.

In choosing to discuss only the practice of the technique, Churchill opted not to interpret it within any established system of medical theory. This decision, seconded by Jukes, not to speculate, not to offer any explanation of acupuncture's empirical successes raises questions about the nature of their response to the technique, and about their accuracy in judging the tastes and susceptibilities of their audience. To a significant degree, Churchill must be considered a good judge of his peers; his *On Acupuncturation* provoked a strong response in the British medical press, and sparked interest in the treatment across a wide spectrum of the medical public. It was reviewed and cited, sometimes with other texts on acupuncture, in the medical press, and these reviews were accompanied by a spate of case studies. Like the book, this reaction was punctuated by passing references to the exotic and the novel; however, the response to Churchill's book was concerned with theory as well as practice – although the two concerns were rarely addressed simultaneously. Similarly, the tone of commentators on the book makes it clear that Churchill's text was well targeted. His decisions neither to dwell on its exotic origins and reportedly miraculous powers, nor to speculate on how acupuncture produced cures were described as 'judicious and unassuming'.[68] The tone of moderation, scepticism and scientific empiricism set by Churchill's opening words – 'There exists a desire for speculation and discovery on the one hand, regulated and qualified by a moderate and proper degree of scepticism on the other' – echoes through the commentaries.

The wave of interest following upon Churchill's publication, however, can only partially be attributed to the persuasiveness or content of his monograph. That the level of interest in the technique had been increasing (albeit from a very low starting point, and in reaction to the European response) was indicated by the two 1820 notes on the subject. In the contemporary literature, there were also references to earlier acupuncture practitioners, notably in Dumfriesshire, Scotland. More generally, French clinical and scientific practices were gaining adherents in the British profession, making Churchill's use of French sources and investigations far more effective than it would have been in 1810 or 1820.[69] Nonetheless, *On Acupuncturation* seems to have provided the catalyst. Acupuncture, if not yet mainstream, at least seemed poised to become a part of the medical discourse in late 1822. References to the technique appeared in the medical press, in some general periodicals, in medical encyclopedia and textbooks, and in the published case reports of several London hospitals.

The medical periodicals presented three basic types of articles about acupuncture. Of these, case studies were by far the most frequent, and can be divided into the categories of foreign and domestic reports. The journals also published the results of experiments investigating the physiological properties and effects of acupuncture, and broader articles essentially reviewing and examining the existing literature on acupuncture, usually with an aim to critique and clarify the therapy and its place in British practice. These latter are the richest source of information about the tenor of acupuncture's reception, and offer the clearest picture of the processes by which acupuncture was judged, assimilated, or rejected by the diverse medical communities of Britain.[70]

'Any means, however ridiculous, which tends to alleviate human misery...': analysing acupuncture, 1822–30

In March 1822, less than six months after its publication, the *Treatise on Acupuncturation* was the subject of an analytical review in the *London Medical Repository*. The article began by citing the earliest European sources of information about acupuncture, noting that these texts recommended its use in gout and rheumatism as early as the seventeenth century. This focus on the European past of acupuncture was a tactic shared by most contemporary discussions of the therapy. However, the *London Medical Repository* article made no attempt to explain

the long delay between these early publications and the eventual introduction of the technique into British practice. Typically, this rhetorical
problem was addressed either by a similar interpretive silence or with an
extenuating response along the lines of Churchill's critique of hyper-
bole. In this case, the reviewer moved immediately from the romance of
its employment 'in countries situated far beyond the Ganges' to a generally positive, although not wholeheartedly encouraging assessment of
acupuncture's introduction and use:

> We consider the present attempt of Mr Churchill to introduce acu-
> puncturation into British practice deserving of our notice, both from
> the judicious and unassuming manner in which the endeavor is
> made, and because we really do think that such an operation may
> be serviceable in the affections against which it has been employed in
> Eastern practice.[71]

The author noted that 'The relief afforded by the operation was both
immediate and permanent.'[72] But the heart of his reaction to acupuncture had as much to do with the on-going redefinition of British medicine as with the technique itself: 'We can see no reason for condemning
any particular method of cure until it has actually been tried, and found
inefficacious. The cases which the author has given, are very interesting.'[73] Bluntly, this reviewer was promoting the idea of experiment and
case studies as authoritative, and as the appropriate way to judge the
value of a medical or surgical technique.

Although the review was framed as an analysis, it also contained full
instructions on how to perform acupuncture, including details on how
'the instruments' were improved, and were not merely sewing needles.
Its author assumed his audience to be already familiar with the growing
literature on the subject of acupuncture, and again noted the painlessness of the procedure:

> The instrument that Mr Churchill employs, is the one improved by
> Mr Jukes, and which appears to be the best adapted of any for the
> purpose. The method to be employed is the following: – 'The handle
> of the needle being held between the thumb and fore-finger, and its
> point brought into contact with the skin covering the part affected, it
> is pressed gently, whilst a rotatory motion is given it by the finger and
> thumb, which gradually insinuates it into the part; and by continu-
> ing this rolling the needle penetrates to any depth with facility and
> ease. The operator should now and then stop to ask if the patient is

relieved; and the needle should always be allowed to remain five or six minutes before it is withdrawn.' This mode of introducing the instrument neither produces haemorrage nor pain; for the fibres are rather separated than divided by the passing of the needle. The introduction of more than one is seldom requisite.[74]

One part of these instructions explicitly involved consulting the patient about his or her unique response to needling (and in a way that anticipated almost immediate relief). This dependence on patient input would play a crucial role in acupuncture's British tenure. The transition to a entirely external reading of the sick body was incomplete in 1822, but it had begun – and yet acupuncture as it was being performed in the West relied on patient self-reporting.[75]

The *London Medical Repository* review, the first available, incorporated Churchill's emphasis on practice over theory, noting without comment Churchill's decision not to offer a theory respecting the rationale of the remedy. The author did offer his own interpretation – one which attempted to integrate acupuncture's effects into western scientific physiology.

> There appears to be but one way in which it can act; and which may be explained conformably to the theories of those very learned and ingenious physiologists, K. Sprengel, and G. Prochaska, which refers those diseases to certain states of the galvanic influence, present in the nervous and muscular fibres, different from that necessary to the healthy function of the part.[76]

Although he did not recite the scientific pedigree of acupuncture, it seems clear that acupuncture's experimental frame and Churchill's tone of rational scepticism positively influenced the reviewer's analysis. Acupuncture, as a completely new and extrinsic technique could be explored and evaluated by the new criteria of experimental empiricism, in a way that surgery's established techniques could not be.

Responses to Churchill's book, and to acupuncture as a therapy, continued to appear throughout the decade. In November of 1823, the *Lancet*'s first volume reported on acupuncture. The article repeated the 1802 pattern of a brief contemporary account encouraging experimentation, followed by a lengthy and detailed quotation from a historical source. However, instead of depending on the widely available lay sources for details, the metropolitan *Lancet* could draw on earlier (and rarer) texts. The author used the 1683 *Philosophical Transactions* account of Ten Rhyne's essay as his source, rather than recycling the by-now standard description from the *Universal History*.

> *Acupuncturation.* Much has lately been said of the efficacy of this remedy in various affections; and well-marked cases in which it has been decidedly beneficial, have been published to the world. In rheumatism, trismus, anascara, it has been tried and with success. The facility with which it may be used, leads us to hope that this remedy may meet with a trial from many intelligent practitioners, who may give to the profession a fair and important result of their observations. Acupuncturation is a remedy of very ancient date. In the Philosophical Transactions for 1683, is a notice of a book written by Dr Ryne, in which an account is given of . . . the various means that the Japanese made use of at that time in the treatment of their complaints, among which is Acupuncturation. [77]

The author's diligence in seeking out this source could be ascribed either to a greater interest in, or a greater access to, the early material. Alternatively, it can be read as an attempt to strengthen the credibility of acupuncture by establishing it in the European medical tradition. The article continued with a lengthy quotation from the *Transactions*, minutely describing the acupuncture needles and apparatus. However, the anatomy-based language in which Ten Rhyne couched his description of the acupuncture body-map (itself already reduced in the *Transactions* to a brief mention of normative insertion points) allowed the *Lancet* version to thoroughly conflate the prescriptive functions of that map with the protective goals of surgical body-mapping in the West. Ten Rhyne's specific points had, in the *Transactions*, become 'the proper places' to needle; in the *Lancet*, this language was taken to its logical anatomical conclusion: 'The chirurgeons keep by them images, wherein all the places in the body proper for the needle, are designed by marks.'[78] In this model, the maps no longer indicated specific points with proven healing effects, but served as surgical chapbooks for an anatomically illiterate profession.

Although the format for this article paralleled that of Coley's 1802 essay, its emphasis – and especially the relative weight given to different types of evidence about acupuncture – was markedly different. The introduction cited well-marked cases and the quotation, though ancient and in language already archaic, was from a medical source far more directly related to the original accounts of the practice than had been the case in Coley's discussion. Moreover, that quote was interpreted and presented in accordance with the anatomical tenets of contemporary surgery. Significantly, the author of this account was careful to close on a note appropriate to the empiricism of the new clinical medicine: '*The author himself was an eye witness of the use of this puncture* on a souldier [sic], who, being afflicted with violent disorders of the stomach, and frequent vomitings at sea, suddenly relieved himself by pricking a thumb's-breadth deep into four different places about the region of his *pilorus*.'[79]

The questions of what precisely acupuncture was – what operations with needles could be called 'acupuncture' and what ailments it could treat – and of how acupuncture produced its healing effects were much discussed. The preliminary articles describing acupuncture were essentially optimistic, perhaps because the sceptics initially expected acupuncture to sink beneath the weight of its own exotic mystery. However, few of these British analysts were willing to take as firm a stand about acupuncture's mode of action as the prompt reviewer cited above. Far more typical were readings which either ignored the issue or drew no conclusions about how acupuncture produced its therapeutic effects. An article in the former category, published by *The Medical and Physical Journal*, uniquely illustrated the frankly social construction of medical credibility as it took place in the case of acupuncture. Its author referred to a set of recently published case studies and to Churchill's anticipated second monograph on acupuncture, 'a body of evidence which shall dissipate the most obstinate scepticism'. He noted, however, that the *Medical and Physical Journal* did not share that residual scepticism: 'For our own parts, we are not at all sceptical on this subject: we are fully sensible that the operation has been followed by immediate and permanent relief in many instances, particularly in that of a nobleman of high rank in the county of Sussex, and who has contributed very largely to extend its reputation, and to enlarge the sphere of its practice.'[80] The social status of the client in this case was shared by the technique he promoted. But the profession was bent on greater freedom from its patrons than this statement suggests. The author concluded with an attempt both to support acupuncture against occasional failures and to bolster professional authority as based on reasoned observation:

[W]e are also aware of its having been employed in vain in fully as many cases... We do not at all intend to depreciate the utility of this simple and easy remedy by this statement, but... to put our brethren in possession of the per-contra side of the account – lest, if they should meet with a disappointment upon their first occasion of their applying the needles, they may hastily and rashly abandon the practice, as supposing it to be founded in delusion.[81]

The fragility of acupuncture, arising from its definition solely through practice and its dependence on exclusively empirical success for support, made it especially vulnerable to rejection if needling failed of its curative effect. No theory existed to validate its cures, or to explain its failures. Proponents of acupuncture were clearly swift to realize this weakness in Churchill's strategy.

From 1826 on, the *Lancet* entered the fray on the side of acupuncture. In their scathing criticisms of acupuncture's detractors, the *Lancet* articles illustrate both how acupuncture was seen by its doubters, and how their arguments were countered by the needle's more strident supporters. One revealing article described the conservative reaction of the Edinburgh medical establishment:

> Acupuncturation has been recently performed in the wards of the Royal Infirmary for the first time during the last session. The subject of the experiment was a man affected with a severe form of sciatica, which had resisted...a variety of applications. The needle was inserted three times, and during, and for some time after each introduction, he was, or fancied himself, considerably relieved. The pain however returning...the idea of prosecuting the practice further was of course abandoned...[82]

Three things are evident in this passage: first, acupuncture was used experimentally only after standard treatments had already failed. Second, the patient's testimony of relief was not, in the case of this exotic therapy, considered conclusive evidence of effect.[83] Third, acupuncture was assessed with greater rigour, and – in light of the patient's previous exposure to more conventional 'applications' – rejected with greater speed than established techniques, especially surgical ones. This is a potent and revealing combination, demonstrating changing values within the British medical profession, as well as the paradoxical nature of its response to the needle. The *Lancet* ascribed this hostile response to acupuncture not to its Chinese origins, but to its close association with France:

> It seems indeed to have been received with indifference by the profession of Edinburgh, many of whom identify it with one of those bubbles which the effervescence of French invention casts upon the stream of public opinion, where floating awhile on the surface,

admired by the crowd, it at length bursts, and 'vanishes into thin air'. Dr Graham [a Professor at Edinburgh]... even ventured a jest on this singular addition to modern practice, by assimilating its application to the conversion of 'the seat of honour' into a pincushion.[84]

The *Lancet* was by no means agreeable to this jocular and casual assessment of acupuncture, and treated Graham's response almost as a breach of professional etiquette, proclaiming: 'any means, however ridiculous in its nature or use, which tends to alleviate human misery deserves a more impartial trial than dismission [sic] into a joke'.[85] The author, pseudonymously named Scotus, assured his readers that acupuncture 'though sometimes uncertain in its results, is of parallel importance and utility with the multitude of medical agents, which are enumerated in every system of surgery and of medicine'.[86] But Scotus asserted that acupuncture had another more powerful claim on the attention of medical men, beyond even its efficacy:

[T]here is besides, something involved in this remedy... The relief in those instances in which it happens is so obvious and instantaneous that curiosity is intensely excited to discover the manner in which such sensible effects are produced. The attempts to explain this mystery... have been extremely numerous and various. Of these, irritation, electricity, oxidation of the needles, and the mental emotions produced by their application appeared the most feasible explanations of that difficulty; but they have each in turn yielded to the opposition of contradictory facts...[87]

Scotus then described the evidence offered against each of these theories. He concluded with a discussion of the evidence against 'mental impressions' as the explanation of acupuncture's effects, but with apparently little hope of convincing his audience: 'Notwithstanding... the obvious disparity between the power of a mental impression... and the extraordinary results of this operation, this view of the subject is beginning to become fashionable amongst the profession.'[88] He argued that the rarity of cures produced by mental impressions, considering how often they might be expected 'completely destroys all analogy between them and the almost invariable effects produced by acupuncture'.[89] He also noted that one objector to acupuncture dismissed the French clinical success of acupuncture as due to 'the greater susceptibility of the people of that country to intense feeling, and despairs of their ever producing much good in the phlegmatic and philosophical inhabitants of the North'.[90] Even French authorities were suspect in the North, despite the strong ties between the medical communities of Edinburgh and Paris. This article and others like it suggest that opponents of acupuncture were no longer passively waiting for it to fade away, while at least some of the needle's supporters were using it as a way of demonstrating the failings of the elites. But beyond these explicitly opposed factions, acupuncture's period as an apparently growing and thriving transplant was coming to a close, at least among those physicians and surgeons who were interested in medical theory as a tool by which to reform the profession and enhance its authority.

As acupuncture's initial prominence in the medical press decreased, frustration with its enigmas surfaced. The *Edinburgh Medical and Surgical Journal* finally joined the discussion with a lengthy and detailed review of recent monographs on acupuncture. All the texts reviewed were foreign; nonetheless, the author of the review article clearly considered acupuncture as sufficiently established in Britain that the journal's previous reticence on the subject required explanation. Like Churchill in his remarks about Chinese 'fables', the editor expressed great scepticism of the original reports about the technique: 'The first accounts of the virtues of the new remedy were so marvelous, and therefore seemed to savour so much of quackery, that, coming as they did, from persons not of the highest authority, we could not but follow the general example, and decline giving implicit credit for their assumptions.' [91]

However, he explained, observations made 'at many continental schools of eminence' meant that the Journal could at long last discuss the topic legitimately. Moreover, the observations of the many European investigators could be compared, and thus formed, in and of themselves, a broader experiment validating the technique: 'the several accounts given by unconnected writers agree remarkably in every essential particular; the alleged facts have been put to the test of a full and minute train of experience by one of the most scientific of the Parisian physicians, in a great public hospital, and under the eyes of its pupils'.[92] Nonetheless, this author presented himself as still doubtful. He agreed that acupuncture's successes could not result from outright fraud on part of its supporters, but he remained disturbed by the mystery of its action. No 'rational way of accounting for its effects' had been discovered despite the long and detailed investigations he had earlier described:

and what is perhaps of more consequence, they have been unable to detect any physiological change...co-ordinate with its operation.

There is in short a total want of every sort of evidence in its favour as a remedy, except that most treacherous kind, the evidence of succession.[93]

Under these circumstances, the author argued that 'a philosophical mind' must discard acupuncture as a placebo – especially because of the sorts of diseases in which it succeeded. Logically, reasonably, the medical community must 'sentence acupuncture to banishment from regular practice, as being nothing else than a variety of animal magnetism'.[94] But how was the evidence presented over almost a decade, showing acupuncture as a 'powerful remedy' for neuralgia, rheumatism, muscular spasms and as also effective in gout, palsy, ophthalmy, pleurisy and erysipelas to be explained? Unlike Scotus, this author concluded by calling for more information on acupuncture's effects on the mind, saying that none of the authors writing about the technique had systematically studied the mental states of their patients.[95]

It is intriguing that by this point, acupuncture's Asian origin, although mentioned in the context of the standard pedigree, was less important than its experimental French rebirth. This author clearly had more information about how acupuncture functioned and was taught in Asia than his predecessors at the beginning of the decade; he noted that there were traditionally 337 spots associated with needling, that diseases were treated by puncture of specific sites among these, and described the examination process of acupuncturists in China. Nonetheless, the article discussed only the most materialistic interpretation of Asian medical theory: 'The Japanese entertain ... the notion that all diseases spring from the presence and accumulation of certain airs in different parts of the body; a very convenient theory in regard to the operation of acupuncture, the needles being supposed to act simply by letting these airs out.'[96] The author immediately dismissed this straw-man, and concentrated instead on the theories of mental impressions, electricity and irritation.

The reviewer focused on France, where acupuncture had been 'a neglected discovery' until a dozen years previously.[97] The individuals he named as its rediscoverers were all experimentalists; like Churchill, he cited Berlioz's 1811 *Mémoire* to the Parisian Society of Medicine, and the Society's subsequent dismissal of Berlioz as 'a rash experimentalist'. He credited Cloquet with the first adequate experiments, by merit of his situation at the great public hospital, St Louis: 'it is to M. Cloquet alone that the credit is due of having first instituted a series of experiments of sufficient extent to permit the practitioner to make up his mind regarding the real merits of the practice'.[98]

Yet although the clinical studies proved the efficacy of the needles, the attempts to clarify acupuncture's mode of action through science and experiment had not been altogether successful. The major results of this experimentation, as this reviewer presents them, were negative:

> [I]t is found, that the rapidity or perfection of the cure bears no reference either to the extent or time of appearance of the red aureole, or to the sensation felt at the moment of the introduction of the needles or soon afterwards, or to the occurrence or degree of oxidation, or to the transmission of electrical current . . . [O]n the whole, they tend to show, either that it operates through the mind, or that it acts by withdrawing a morbid accumulation, or rectifying an aberration of the nervous fluid.[99]

In other words, they eliminated few theories and confirmed none, while failing even to answer empirical questions about practice, such as how long to leave the needle *in situ*. More importantly, the experiments indicated no method by which to determine whether or not the needle was having an effect other than through the patient's self-reported response.

The review assessed the developments of the three major theories, noting how each failed, and how each nonetheless offered some assistance to understanding and rationalizing acupuncture's curious effects. For example, in the case of the electrical explanations, the author noted,

> [A]lthough the electric theory cannot stand the test of examination in the shape in which it was thus originally conceived, it must be admitted to be very difficult to account for the effects of acupuncture without having recourse either to the properties of the electric fluid, or to those properties of the nervous fluid, by which it is associated with electricity.[100]

The reviewer described himself as forced to accept, for lack of a better explanation, a theory based on new ideas of the relationship between electricity and nervous action:

> It is by uniting these [nerve] currents more directly than the nervous organization admits of, or by diffusing and moderating local currents of preternatural force or quickness, that . . . the acupuncture needles operate . . . [V]arious nervo-electric currents must really pass through the parts into which the needles are inserted, and then needles, as

being good conductors unite them together. For... the annular oxidation so universally remarked on the needles can arise from nothing else than the union of so many opposite currents...[101]

In fact, this theory called on the latest evidence on the behaviour of both the nerves and electricity in predicting that the needle acted as a switch or preferential channel for the body's equivalent of the galvanic fluid.

Perhaps because of the uncomfortable novelty and experimental intractability of his hypothesis, the author then returned to the tempting theory of the mental action of the needles: 'It is certainly very natural to expect that the operation of acupuncture should be attended with peculiar impressions on the mind.' Damningly, he added that the 'formality and apparent mysticism' in the performance of therapeutic needling was surely sufficient 'to inspire the patient with a portion of that confidence, which forms the cause and sine qua non of success in the case of animal magnetism'.[102] Churchill's attempt to standardize the practice of acupuncture while maintaining its wide therapeutic appeal was seen by this contemporary as dangerously close to the unexplained rituals of mesmerism. The writer was candid enough to admit that he was, simply, reluctant to accept acupuncture however it was supported:

> such is the unwillingness of most people to admit the remedial virtues of acupuncture at all, and such, it may be added, is our own hesitation to admit them to the extent which the argument now used implies, that – without any special reason, without any other reason, in short, than the known uncertainty of medical facts when unconfirmed by repeated experience – it is exceedingly difficult to grant the premise, on account of the magnitude and singularity of the conclusions.[103]

This residual unwillingness seems to have derived from acupuncture's intractability to theory. No satisfying explanation of its success could be found, and those successes were therefore suspect. Nonetheless, the author noted that 'the permanency of the cure... is apparently irreconcilable not only with the effects of confidence in an imaginary remedy, but likewise with the therapeutic effects of all mental impressions whatsoever'.[104] Empirical results still had some authority, and cures, whether explicable or mysterious, were still valued by both practitioner and patients. Moreover, he considered Cloquet, his major source, to be reliably sceptical (despite the misfortune of his French birth). Even those traits so valued by Churchill – simplicity, efficacy and celerity – could become reasons to distrust cures produced by the needle. All former methods of cure were both complex and uncertain; how could this mysterious and foreign cure be so much more effective than these fruits of European medical expertise and natural philosophy? As the reviewer noted, 'M. Cloquet . . . looks upon it with a degree of distrust . . . such as a philosophic mind cannot easily lose sight of, on comparing its simplicity, efficacy, and celerity, with the complexity, uncertainty, and tediousness of the former methods of cure.'[105] Simplicity and safety, while admirable, were not seen as transparent evidence of greater merit, or grounds to discard therapies that could be explained within the western medical tradition. With more resignation than delight, the reviewer brought his report to its close:

> [I]t is impossible to deny a great part of the facts which have been mentioned, without refusing credit to human testimony altogether. In the foregoing abstract, an attempt has been made to appreciate the truth and value of the works analysed, by the evidence which they themselves afford . . . It is not unreasonable to insist, that those who are inclined to sneer at acupuncture, of whom there are not a few, will submit to be guided by the same principle.[106]

The reviewer was hardly ardent; rather he presented himself as inescapably forced to credit the incredible, at least to the extent of putting acupuncture forward for yet more trials.

The depth and extent of this analytical review make it clear that even after a decade of study, use and diffusion, no available theory could easily encompass the complete range of phenomena presented by acupuncture. Earlier in the decade, the journals had still hoped for some kind of scientific clarification of acupuncture's mode of action, and one genre of writing on the treatment had been devoted to reporting European experimentation on acupuncture's physiology. In 1825, several journals published accounts of French experiments on the 'electro-magnetic' or 'galvanic' phenomena related to acupuncture. They did not achieve consensus about the results, or even about the appropriate experiments to perform. However, the assays were described in some detail, and employed surprisingly sophisticated methodologies. All were dedicated to discovering whether 'the electrical fluid' was the cause of acupuncture's healing properties. Certainly, each of the French experimentalists had discovered that inserting needles into living bodies would produce electrical effects measurable by their equipment. However, the relationship of those effects to the effect of acupuncture on the sick body was made no clearer. The *Medical and Physical Journal* provided a vivid ex-

ample of this confusion. Its report stated clearly that 'perceptible quan-
tities of this [galvanic] fluid are always disengaged by a needle plunged
into a part of the human body affected with pain'.[107] And yet, the
experimenter himself had concluded that acupuncture could not be
explained by galvanism:

> *M. Pelletan thinks these galvanic phenomena unconnected with the curat-
> ive effects of the operation; which opinion is founded upon the circumstance
> of the relief being in no case in proportion to the quantity of the fluid
> disengaged; and that very marked effects result from the acupuncturation,
> even with a needle terminated by a non-conductor*...[H]e infers that,
> in the cases alluded to, the effects were the result of the oxidation
> of the metal.[108]

Nonetheless, both Pelletan and his English interpreter insisted upon the
'incontestable success follow[ing] the use of the acupuncturation, particularly in rheumatic affections'.[109] A second article, without describing the experiments in any detail, concurred with the conclusion that
galvanism could not explain acupuncture.[110] However, a third article,
'On the Electro-Magnetic Phenomena Observed in Acupuncture', came
to the opposite determination, and encouraged users of acupuncture to
choose their needles wisely, ensuring that they were made of reactive
metals.[111] This series of conclusive but contradictory results would certainly not have inspired potential users of the technique, much less
pleased those of its promoters who were attempting to explain and
theorize it persuasively in the context of the new medical science.

However, not all of acupuncture's supporters committed themselves
to discovering the therapeutic mechanisms behind acupuncture's successes. In the numerous case studies, whether extracted from foreign
journals or published as original communications, the opposite trend
was visible. Few of these users were even interested in commenting on
the debate over acupuncture's mode of action, much less in offering an
opinion on the subject.[112] Instead, they offered the details of each
application, discussions of the needles or other apparatus used in their
versions of the acupuncture operation, and strictures on which ailments
are appropriately treated with the needles. They also modified the technique. Their authors often showed great enthusiasm for acupuncture. In
1823, Frederic Finch, a surgeon in Greenwich, declared:

> Having in the course of the last month, resorted to acupuncturation,
> with manifest advantage, in a case of fixed pain in the lumbar region,

> I am therefore persuaded that it is a means of relief which deserves to
> be extensively known and generally adopted... The Profession must
> feel much indebted to Mr. Churchill for bringing this ingenious and
> valuable practice into notice.[113]

Finch then described as his own innovation the use of needle-pricks to
provide outlets of oedematous fluid without scarring, calling this too
'acupuncturation'. Indeed, the advent of acupuncture and its novel use
of needles to treat a wide range of ailments seems to have triggered a
reconsideration of the tool in surgical practice. Later that year, Finch
published a second case in which he had tried acupuncture for tetanus.
Here, his inclination to experiment received encouragement not from
theory but from experience and analogy; he had previously used needling successfully to treat rheumatic rigidity in the muscles, and thought
that acupuncture might alleviate the stiffness of tetanus.[114]

Another surgeon followed a similar path to the use of acupuncture,
and became a similarly ardent supporter: 'Having witnessed the instantaneous, and I may add, astonishing effect of Acupuncturation as a
remedial means in the first case that came under my notice, I resolved
to give the operation a fair trial in every instance wherein its use should
be indicated.'[115] T. W. Wansbrough, the author of that endorsement,
was also careful to indicate who could expect such successes in the
practice of acupuncture – only 'in the hands of a skillful surgeon'
would acupuncture show its power. Rather more contentiously (given
the medical politics of the time), Wansbrough also asserted that only to
'a *liberal* mind' would this 'convey the pleasurable gratification of
affording to human suffering instantaneous relief, by a more expedi-
tious and efficacious mode than any other remedy in such cases hitherto
known and employed'.[116]

Liberal or not, Finch and Wansbrough were not alone, either in their
enthusiasm or in believing that 'acupuncturation may be employed in
various ways in surgery with advantage'.[117] By 1823 Harry Carter, Senior
Physician of the Kent and Canterbury Hospital, was also using 'acupuncture' in both of its major forms. Carter declared 'I shall employ acupunctura, which I have found... safe and effectual' in oedematous cases,
while in rheumatism and sciatica 'acupuncture was of decided
efficacy... [and] put the disease to flight.' His use of two different
terms suggests that Carter differentiated between mechanical and
analgesic therapeutic needling.[118] And of course, Carter returned to the
theme of safety. Judging by the number of times proponents mentioned
safety, this aspect of acupuncture's practice was more important to

practitioners than anything else, including how the treatment attained its results. Acupuncture's low risk was clearly seen as one of its most attractive and persuasive features. Finch summed up the reasons for promoting acupuncture quite bluntly: '[I]t effects its purpose in the most easy manner, and should it fail, is productive of little or no inconvenience or uneasiness to the patient.'[119] Unlike bleeding or administering purges, mercury or leeches, acupuncture didn't hurt, and wouldn't scare the patients away, even if it failed to heal them. Wansbrough (after inciting 'timid practitioners' to try acupuncture) did note the questions about acupuncture's mode of action. However, he was only interested in critiquing the 'Chinese' model of its activity as interpreted by Ten Rhyne and Kaempfer, and did not offer one of his own.[120] He was patently unconcerned by the ongoing debates, concluding blithely, 'I believe it to depend on those mysterious operations of nature that will ever be beyond the reach of human ken, and which by consequence constitute the *ne plus ultra* of physiological research.'[121] In an editorial comment following the case studies, the *Lancet* gently mocked Wansbrough's enthusiasm, his assertions of 'the magical effects of the needles' and his military metaphor – 'Doubtless, the author receives all the ''blessings'' and thanks which successful generals are wont to receive. We only doubt, whether he ought not to have a corona muralis. His despatches are interesting.'[122] Clearly his decision not to theorize was unremarked and unremarkable.

Acupuncture established?

> Without the fostering care of a great name, my prophecy has been verified, and acupuncture is now employed, not only in the Eastern Hemisphere, in France, and in America, but throughout the British Dominions, and in our London Hospitals, under the auspices of men, who stand deservedly high in the ranks of literature and science.[123]
>
> Churchill, 1828

Throughout the decade, British authors of acupuncture case studies consistently noted that the technique remained virtually unknown, or at least undervalued in Britain. Yet by the end of the decade, the term did not need to be defined or explained, and journals no longer felt it necessary to explain their inclusion on articles on the technique. So how thoroughly domesticated was acupuncture? Was it at all integrated into the surgical canon of the times? In addressing these questions, it is important to look beyond the representations and discussions of acupuncture in the medical periodicals. Despite their importance in disseminating innovations and imports, the impact of the medical journals was ephemeral. They addressed a wide variety of different audiences, but they were not primarily intended to instruct the next generation of practitioners, nor – at least by the nineteenth century – to speak to the wider public. And they were not necessarily designed as a lasting information resource.[124] Medical encyclopedia and compendia were intended both to instruct and to endure; if a technique was attractively described in one of these resources, it gained a greater chance of surviving the shift of medical generations. Information about acupuncture did appear in a distorted form in such sources during the eighteenth century, and by the 1830s the technique had an established place in this genre of medical literature.

The *Cyclopedia of Practical Medicine*, a product of the tensions between medical radicalism and conservatism, offered a substantial body of information on acupuncture in 1833.[125] This article was written by University College London's young Professor of Practical Medicine, John Elliotson, an avid borrower of European medical novelties with a growing reputation for medical innovation and political and scientific radicalism. Elliotson initially contrasted the purely mechanical form of 'acupuncture' with its history and practice in China and Japan:

> The most obvious purpose of this operation, is to allow the escape through the skin of the fluid of oedema or anascara; or of the blood when superficially accumulated; but – from an idea that various disorders arise from a subtle and acrid vapour pent up – it has been had recourse to by the Chinese, for the purpose of giving vent to this vapour, from time immemorial.[126]

Throughout this section, Elliotson persistently interpreted Chinese theories of acupuncture as referring to a physical accumulation of gas. Moreover, he provided the exotic details which had disappeared from the medical periodicals – golden needles and ivory hammers are lovingly described, as are the existence of acupuncture points, and the importance ascribed to them by Asian practitioners. Elliotson also described the path by which acupuncture became known to western medics, reciting the conventional sources, from Ten Rhyne to DuJardin and Vicq D'Azyr. Unlike his journalistic colleagues, Elliotson offered an explanation for, and critique of, European disinterest in the therapy:

Owing partly to the frightfulness of running needles into the flesh, and the high improbability of any benefit from such a practice, a hundred and seventeen years elapsed before any European practitioner made trial of it. Dujardin, in his 'Histoire de la Chirurgie,' and Vicq D'Azyr, in the 'Encyclopedie Methodique,' mentioned it about a century after Ten Rhyne had published; but only to congratulate the world that the statements of the latter, and of Koempfer, had not induced anyone to practise it. The first European trials were made by Dr Berlioz of Paris, in 1810. Its power proved so extraordinary, that he employed it very extensively; and numerous French practitioners imitated his example with the same results. A body of similar English testimony followed; and acupuncture affords a striking instance of a good remedy discovered from a groundless hypothesis, and condemned, without a single trial, for more than a century.[127]

With this description of acupuncture's arrival and scientific validation, Elliotson's narrative changed; after the breathless exoticism of the introduction, it suddenly became pragmatic, even directive in tone. The article laid out precisely which ailments acupuncture had successfully treated, and quantified acupuncture's merits numerically, in terms of patients cured.[128] In this entry, Elliotson included instructions on acupuncture practice, and comments on how safe and painless the operation was. He concluded it by concisely examining and dismissing each contemporary explanation of acupuncture's mode of action, opening with the obvious: 'The modus operandi of acupuncture is unknown.'[129] The form of Elliotson's article, and its contradictory emphases on historical exoticism and contemporary pragmatism were typical of longer discussions of acupuncture. Alternatively, authors mentioned and described acupuncture in the context of the illnesses for which it was indicated, or with a group of similar therapies. This style of presentation closely paralleled the way in which acupuncture and other new techniques were taught to medical students, at least by Elliotson in his lectures at St Thomas's Hospital and University College London.

In terms of gauging the breadth and depth of the interest in acupuncture, and of the British response, two factors should be considered beyond the frequency, length and tone of articles published on the subject. First, British case studies were often published more than once; two or more journals might pick up a case study and reprint it, with or without citing the original source. Churchill and Elliotson frequently cannibalized their own and other supporters' published materials in subsequent publications. In fact, Churchill referred to this recycling and re-exposure as a kind of validation of the material re-presented:

> I now proceed to furnish impartial evidence to the truth of what I have advanced, and that it may come before the public in the most unsuspicious light, I shall enumerate but few cases of my own; nor is it my wish to avail myself of all the unpublished ones, with which I have been furnished by their respective authors – abundant materials have been supplied by our medical periodicals...[130]

Second, many of the case studies and all of the experimental studies printed about acupuncture were picked up as part of regular surveys of the foreign medical literature. Clearly, there was sufficient interest in acupuncture to prompt this kind of borrowing; on the other hand, the level of repetition and the dependence on foreign sources argues for a paucity of acceptable submitted material on the subject.[131] At least one periodical, the *London Medical and Physical Journal*, actively sought domestic case studies on acupuncture during this period.

Of course, pragmatism was the primary cause of this interest in acupuncture, and of the positive response to it. The explicit emphasis on safety and efficacy expressed in the case studies, the monographs, and in many of the analytical articles illustrates the importance that this aspect of acupuncture held in the eyes of acupuncture's medical popularizers. In a crowded medical marketplace, an inexpensive, relatively painless, and entirely exclusive technique which could effectively treat the common pains of workers, and the rewarding, but rarer ills of the middle class was not to be scoffed at as long as it worked – or at least did no harm. In the early years of acupuncture's use in Britain, its successes were public, impressive, and readily available for consumption through the medical journals. Foreign and local cures were described in detail in the expanding medical press (which was, in any case, looking for new material with which to fill its pages) and in lectures to London students. Acupuncture was also discussed in monographs, published volumes of lectures, and medical compendia. While some evidence contradicting the rhetoric of safety was published, it was generally presented as exceptional or as an example of improper use of the needle.

In the decade following the publication of Churchill's monograph, acupuncture was presented to the British medical public in greater detail and in a more accessible form than ever before. The explosion of the medical press, and its fragmentation allowed for more views and innovations to gain visibility to a wider range of the medical profession than was imaginable in the eighteenth century. The use of acupuncture was still on the peripheries of regular medicine after a decade of British evaluation and practice; however, its position of precarious normality was significantly different from one on the medical fringe. Unlike mesmerism or patent medicines, neither form of acupuncture was ever excoriated by the medical press. Its most outspoken opponents linked it to animal magnetism, but only implicitly; doctors who supported acupuncture remained within the orthodox medical establishment. Evidently they also either retained or expanded their patient base; Churchill in 1822 was practising in the slightly seedy Leicester Square area, while the *Lancet* in 1829 found him better-situated in Grosvenor Square. Moreover, a diverse range of medical practitioners (although not apparently a large number of them) took up acupuncture. The technique was used both in private practice and in the great London hospitals – and its use was not confined to the metropolis. Acupuncture was taught to medical and surgical students, and had a place in the durable resources of medical practice, the encyclopedias, dictionaries and compendia.

4
Networks and Innovations: the Persistence of British Acupuncture, 1828–90

> ...it is for the doubting and hesitating, to impeach the veracity, or prove the credulity of the many, who have produced, witnessed, or felt the truly beneficial effects of acupuncturation.[1]
>
> Churchill, 1828

In 1823, James Morss Churchill pledged to bring his medical audience an 'appendix' on acupuncture, presenting a collection of cases in which the needles had been successfully deployed. Five years later the book, *Cases Illustrative of the Immediate Effects of Acupuncturation, in Rheumatism, Lumbago, Sciatica, Anomalous Muscular Diseases, And in Dropsy of the Cellular Tissue; Selected from various sources, and intended as an Appendix to the Author's Treatise on the Subject,* finally appeared. As promised, a series of thirty case studies formed the bulk of the text. They depict the everyday use of acupuncture by eleven British medical men, both surgeons and physicians, between 1821 and 1827. Churchill clearly intended the book to be a manual for would-be acupuncture users as well as a repository of evidence supporting the technique. Consequently, the cases were richly detailed, describing the symptoms of each patient and the exact manner in which the needles were placed, inserted and removed. In these minutiae, the voices of the individual practitioners, and occasionally even those of the healed patients, can be heard, making it possible to reconstruct much of the process by which acupuncture diffused through Great Britain, geographically and socially.

Networks and witnesses: persuasion and diffusion beyond the printed page

> I have been induced to try it, merely at the urgent request of the sufferers. I have, however, always been anxious to avoid the

importunities of patients, and merely to employ this valuable agent, in cases that appear adapted for its use.[2]

Churchill, 1828

As the author of Britain's first monograph on acupuncture, Churchill found himself at the centre of a growing network of acupuncture-using practitioners. He drew upon this network, as well as his private practice, to supply cases for his 'appendix', but – in recognition, perhaps of the changing standards of proof in medicine – he also borrowed extensively from the 'abundant materials [that] have been supplied by our medical periodicals'.[3] Churchill and his colleagues used acupuncture on middle-class and wealthy patients as well as on their working-class and charity patients, demonstrating that the needle was no longer considered likely to frighten away paying clients. Indeed, Churchill's clientele included an admiral and member of the House of Commons (incidentally, also an inventor), several gentlewomen, two labourers, a world-traveller, a gentleman of independent means, and one working woman. Those of the other ten practitioners were similarly varied, ranging from an earl to a well-digger to a naval pensioner. Of the thirty sufferers who were described in detail, nine were women, and sixteen were recognizably working-class or poor, while the remainder were either of independent means, wives, or shop-owners. Many of these individuals shared a trait transcending the boundaries of class and gender: they were desperate. Half of the thirty sufferers turned to acupuncture as a last resort; others, while not as despairing, had tried other remedies without success.

The story of Miss Wildman, a milliner treated by Churchill, exemplifies this attitude towards the needle. Her case also illuminates one of the routes by which information about acupuncture diffused into the lay community. In a letter written at Churchill's request, Wildman described her peregrinations in search of a cure for persistent pain following a fall. After enduring massage, moxas, frictions, embrocations, bleeding, and finally electrical treatments in London, she wrote: 'I was recommended as a last resource to try puncturation, which was performed by Mr Churchill with the greatest success.'[4] This recommendation was offered by an established physician at Guy's Hospital in London whose use of electrical stimulation suggests that he was himself moderately progressive. Wildman had apparently never heard of acupuncture before it was recommended to her, and the physician did not himself practise needling. However, he did know of Churchill's success with the implement. For Wildman (and probably for her referring physician) the decision to try acupuncture was born of desperation. Nonetheless,

Churchill's report of her case made it clear that if Wildman considered acupuncture a last resort, she also saw it as an avenue worthy of active pursuit. He wrote that when the milliner first came to him in 1822, 'I considered it a hopeless case, but at her earnest request, introduced needles.'[5] Both expressed some surprise at the instant cure effected by acupuncture. 'Last resort' cases like Wildman's undoubtedly offered dramatic proof of acupuncture's unique powers – the needle not only healed, but healed in cases where no other technique, orthodox or experimental, had been effective – and Churchill almost certainly chose them strategically. But the fact that acupuncture remained a therapy of last resort also suggests that incredulity was an important component of the British response to acupuncture. Indeed, one woman of quality only allowed herself to be needled (at the suggestion of her practitioner) when she had reached the breaking point: 'when acupuncturation [was] proposed, she consented to the operation with this remark, "anything to relieve me from this agony."'[6] When the operation swiftly and completely relieved her facial neuralgia, its success dumbfounded the doctor as well as the patient. In particular, her physician expressed surprise at the needles' unassisted therapeutic power: 'Although I have performed this operation many times, and been present when others performed it, I have never seen a case in which the efficacy was so decided, or in which the relief...was more inquestionably [sic] attributable to the action of the needles.' The patient, meanwhile, refused to allow her doctor to remove the needles, fearing that the pain would immediately return. Clearly, like her working-class counterparts, she remained unpersuaded of the durability of acupuncture cures. Her doctor only gained her consent by persuading her that removal was essential to the needles' continued therapeutic efficacy.

The practitioner's need to persuade and negotiate extended beyond the sickroom, and acupuncture's wealthy lay supporters were fully capable of exploiting their bargaining power to further diffuse the technique. Sir Isaac Coffin (admiral and MP) sent for Churchill to needle his rheumatic hip whenever he was in London. He also insisted that his Sussex practitioner both treat him with acupuncture and send reports of all of his acupuncture cases directly to Churchill, for inclusion in his *Cases*. Lay promoters of acupuncture were often those who most utterly despaired of orthodox medicine, sometimes after years of brutal and ineffective – but theoretically sound – 'regular' medical treatment.[7] The most prominent lay proponent of acupuncture was the third Earl of Egremont, George O'Brien. His enthusiasm for acupuncture was well-known to the wider medical community, and his zeal apparently extended to lobbying the medical press.[8] Thomas Martin, who became the Earl's surgeon, described O'Brien's conversion in a letter to Churchill. He began with no small boast:

> In my hands [acupuncture] has been singularly fortunate, and successful in its effects; . . . it will afford you no small share of gratification to hear of the rapid strides it is making towards popular favour, in consequence of my having introduced it successfully in the case of a noble Lord, residing in this county. . . His Lordship sent for me, having heard of my previous successful cases, through the medium of his noble Daughter, who had witnessed them . . .[9]

George O'Brien, the third Earl, had been confined to his bed for five weeks with sciatica; no established remedy, medical or surgical, had relieved his condition. He had received from his daughter a first-hand account of an acupuncture cure and was desperate enough to send for Martin on the strength of this testimony. After his sudden release from pain, Martin and acupuncture equally soared in his regard: 'There are no bounds to his Lordship's gratitude and delight: he went almost on purpose to Brighton, a distance of thirty miles, to make it known amongst the nobility and faculty there.' Martin went home in his lordship's coach with a fat cheque and the lucrative assurance that his name would ring in the ears of the Earl's acquaintance. Apparently, Churchill himself subsequently treated O'Brien in London, for 'a slight attack of rheumatism', at which time he cannily secured permission to further broadcast the case.[10]

The medical press, and in particular the medical periodicals, had been central to the early diffusion of acupuncture. However, although these journals played an enormous role in the original diffusion of information about acupuncture – for certainly not all of its users in the 1820s and 1830s learned of it directly from Churchill's monograph – by 1828 the practice was clearly being spread through other channels as well. As will be discussed later in this chapter, a far broader audience was exposed to acupuncture through medical education, both as students who would subsequently disperse across the Britsh Isles, and through serialized and collected lectures. Moreover, well-connected patients like Sir Isaac Coffin and the Earl of Egremont communicated their experiences with acupuncture to their equally well-connected acquaintances, extending the credibility of the therapy at the same time. Indeed, several medical journals eventually reported on the phenomenon of elite amateur activism. While some disparaged its results as mere medical modishness, others did regard acupuncture's upper-class support as enhancing its authority.[11]

Poor patients too told their families and friends of the technique which had restored them to health, although the impact of their support was strictly local. T. W. Wansbrough treated a 70-year-old man, troubled with lumbago brought on by years of working outside in all weathers. In talking with this man, Wansbrough learned to his surprise that he had walked nearly a mile specifically 'to be relieved by my needles', about which he had somehow heard. A few days later, another ancient appeared at his door with the same problem, asking for the same treatment – three needles into the muscles of the lower back. He had been sent by Wansbrough's earlier patient. Some time after these two patients had been treated successfully, a third old man, formerly 'a labouring gardener' in the same area, arrived also requesting acupuncture. In seeking professional medical assistance, the men were making an investment; in specifying acupuncture, they were expressing confidence in it, and Wansbrough complied with their requests. These cases indicate that paying clients, even when of a lower social rank than their medical practitioner, were still able to take some control over the therapeutic encounter. Moreover, they illustrate that information about acupuncture had filtered through to the working classes, and suggest the very local but clearly effective means by which this diffusion occurred.[12]

Within the profession as well, local networks were emerging and played a role in the diffusion of information about acupuncture. At the 18 March 1833 meeting of the London Medical Society, after a somewhat rambling discussion of rheumatism, one member of the Society asked his colleagues their opinion of acupuncture as a therapy for that ailment. He mentioned that it had failed in his own 'limited experience'.[13] A fellow surgeon, named Dendy, replied, 'and in mine too. When first it was proposed, it certainly effected some singular cures, but, of late, success does not seem to have attended it.'[14] However, Dendy then regaled his colleagues with a story of one such 'singular cure' involving a surgeon, a nobleman, and a horse named Acupuncture. The story was, of course, yet another version of Martin's successful cure of George O'Brien. Once again, this case offered an argument for acupuncture which was sure to appeal to this company of moderate medical reformers and mid-level practitioners: Dendy preceded his narrative with the note that this was a cure which 'benefited both patient and practitioner in a very agreeable manner'. Dendy also gave a more careful account of the Earl's pre-acupuncture sufferings than had either

Martin or Churchill. He commented in particular on the inability of the established London doctors to cure his ailments – a failure which could not fail to please the moderate reformers who made up his audience:

> The Earl of Egremont was a martyr to rheumatism, and some years since, after having been treated by every medical man of note in London, without obtaining relief, he retired to his seat at Petworth, in despair. A friend of mine [Martin], who resided in Sussex at that time, happened to get an early copy of Mr Churchill's little work on acupuncture, and tried the remedy therein advocated with perfect success on an old woman who was a protégé of Lady Burrell, the daughter-in-law of the Earl. Her ladyship heard of the cure, and told the Earl what had been done; the result was, that the surgeon was sent for forthwith to try the new process on the peer, into whose tortured person he accordingly introduced two needles, keeping them in for twenty minutes. The effect was, that the Earl, who had obtained no sleep for the past fortnight, that night slept for seven or eight hours. Filled with joy, he gave the fortunate practitioner a check for a large sum, sent him home with post horses, and that day bestowed on one of his favourite racers the name of 'Acupuncture'. The event made my friend's fortune.[15]

Dendy's version makes several things clear: first, that no experimenting was done on the Earl. The old woman under Lady Burrell's protection had been Martin's 'clinical material', and only after the success of acupuncture in her case had erased the observers' doubts was the Earl told of it. Similarly, Churchill's first four patients, whose cases he reported in the *Treatise*, were all poor: three were labourers, and the fourth was a woman who had turned to a Poor Law hospital for medical care. Secondly, the Earl actively sought out 'the new process' as an alternative to the failed techniques of establishment medicine. Discouragingly, at least from the perspective of the innovative practitioner, the Earl, like many such disgruntled consumers, directed his subsequent enthusiastic networking towards promoting the technique of acupuncture rather than the individual practitioner who had treated him. The horse, after all, was named 'Acupuncture' and not 'Martin'. Thirdly, Churchill's monograph was the source of Martin's information and provided the template for his practice of needling. But Dendy ended his monologue with the cheerless postscript that 'As regards my own experience, however, I may state that I have lately had three cases in which I have tried this remedy without advantage.' Thus acupuncture's success – and the money it had made for one practitioner – was bracketed by its subsequent failures. Neither Dendy nor his fellow medics offered any suggestions as to why acupuncture might have ceased to be effective (or, for that matter, why it had ever succeeded) and the conversation lapsed.

Like the case studies put forward in Churchill's second monograph, Dendy's story and the convivial setting in which it was told offer clues about both the public disappearance of acupuncture, and its persistence away from the spotlight of the medical periodicals. Local groups like the London Medical Society, moulded equally by the tradition of gentleman's amateurism and by the newer demands of competition and professionalization, provided a social forum within which medical men could discuss and evaluate new developments, and quietly exchange expertise. Along with mesmerism, galvanism, new drugs, the use of the stethoscope and myriad other medical innovations of indeterminate worth, acupuncture was considered and discussed at these social gatherings. Dendy's tale about acupuncture, narrated as acupuncture was disappearing from the printed media of therapeutic practice, reflected the ambivalence of professional responses to the healing needle.

The informal connections forged by individual practitioners also played a vital role in the diffusion of acupuncture. As Dendy told his audience, Thomas Martin had managed to get an 'early copy' of Churchill's *Treatise*, which had in turn 'made his fortune' by enabling him to relieve the aching Earl. Martin's stroke of luck depended on the fact that Churchill and Martin had met and become friends while training at the United Hospitals (Guy's and St Thomas's) in London. Churchill himself had initially learned of acupuncture not from reading the published French reports of it, but from another surgeon, Mr Scott of Westminster. Churchill's interest was aroused by privately communicated and subsequently directly witnessed successes with needling.[16] Scott himself, like Tatam Banks, may have seen acupuncture performed in Paris.[17] Indeed, even before Churchill published his *Treatise*, he was forging links with other British practitioners interested in or using acupuncture. Scott had asked a fellow surgeon to perform acupuncture on him, and to witness and record the results. This second medic had sent the results to Churchill. Indeed, the act of witnessing was a frequent point of contact and potential transmission, and one which also illuminates contemporary structures of authority and its propagation. Churchill and his successors regularly listed the names of prominent observers who were in attendance upon successfully cured acupuncture patients, implying that their presence added weight to the reported results.[18]

Although it has been argued that acupuncture was, from its introduction, a fringe or 'quack' medical practice, its initial proponents in nineteenth-century Britain (as earlier in France) were by no means considered radicals or quacks – or at least were not so regarded because of their use of acupuncture.[19] Instead, most of its proponents were precisely the moderate medical reformers and mid-range GPs and surgeons who formed the bulk of the medical public in the first half of the nineteenth century. Thus, the London Medical Society, where the Earl of Egremont's case had been so vividly retold, was publicly critical of the Royal Colleges but was not marked by the level of radicalism characteristic of the Westminster Medical Society (where discussions focused on highly politicized techniques like mesmerism rather than on apparently atheoretical – and hence almost apolitical – innovations like acupuncture). Only a scattering of medical 'prominents' leavened this mass – including John Elliotson (1791–1868) and James Wardrop (1782–1869) – and they shone in London hospitals and Court bedchambers as well as in radical circles. The former had been educated at Cambridge as well as Edinburgh and Guy's Hospital; became a Fellow of the Royal College of Physicians at the age of 31; held the Chair of Medicine at the new London University, and was Physician to St Thomas's Hospital. But Elliotson was also a founding member of the London Phrenological Society and the London Mesmeric Society. It was his involvement with mesmerism, rather than his frequent employment of the needle in the wards of St Thomas's Hospital, that forced Elliotson to resign his positions and found the London Mesmeric Infirmary. James Wardrop was Surgeon Extraordinary to George IV during his Regency, then the official Surgeon during his reign. His training was far less exalted than Elliotson's; he apprenticed with his surgeon uncle, then toured the hospitals of London and Paris. Wardrop sided with the radicals in the struggle for medical reform, writing the scathingly satirical 'Intercepted Letters' column for the *Lancet*, and eventually refusing a baronetcy for his services to the king. Nonetheless, he died a Fellow of the Royal Colleges of Surgery of both London and Edinburgh.[20]

As was typical of the membership of provincial, suburban and metropolitan medical societies, most of the men who publicly took up acupuncture depended on their practices for both income and for 'clinical material'. Churchill noted that he could not perform the experiments necessary to establish acupuncture's active principle because of the small size of his practice, and complained about his lack of access to the hospitals.[21] Other proponents of needling reported on single cases even after multiple case studies had become the norm, precisely because they saw little chance of getting another in their limited practices. A surgeon in Ayr, reporting late in the century on a solitary case in which acupuncture had relieved the pain of a man dying from cancer, prefaced his data with the apologetic acknowledgement:

> [O]ne case goes only a short way in establishing any method of alleviating or curing the pain of this formidable disease, but a long interval may pass before another presents itself in a small provincial town with a sparse surrounding population. Hence my reason for publishing a single case.[22]

As the reconstruction of medicine along scientific lines proceeded apace, the power of individual practitioners without hospital or dispensary connections to influence or inform medical practice diminished. This may have harmed acupuncture's prospects significantly. Moreover, in private practice, patient satisfaction remained crucial; doctors had not yet established themselves as the sole interpreters of the body, or as purveyors of exclusive insights into health and disease. Laymen and women, as well as midwives and unorthodox practitioners, still claimed to be authoritative observers of the body, demanding (and receiving) treatments based on their self-diagnoses.[23] The difficulty which most of acupuncture's professional users had in performing experiments either on acupuncture or with it, combined with the changing standards of medical journalism, partially explain acupuncture's near invisibility in the medical press between 1835 and 1870.

The declining visibility of acupuncture, 1828–70

> When I told him I could, I thought, relieve him in five minutes without his taking a drop of medicine, he imagined I was joking and replied, 'you'll be more successful than I take you for then.' He did not, in the smallest degree object to my using the needles...[24]
>
> John Tatam Banks, 1831

Several of Churchill's case studies chronicle patients responding to their operations in ways that indicate a curious disbelief of their own senses; these sufferers react as if they cannot credit their own experiences of complete relief from painful symptoms. For example, a cellarman named Field felt no pain after his acupuncture operation, but 'he was sceptical as to its having removed the disorder, for his first attempt to move, after the needles were withdrawn, was made with the greatest caution; and when he found that he was really freed from the disease, he could not divest himself of the fear that it would immediately recur'.[25] The simple, painless – and, since he was needled in the loin while lying flat, invisible – operation of acupuncture clearly did not fulfil Field's expectations of an operation, despite its therapeutic efficacy. Moreover, like so many acupuncture patients, Field had previously been disappointed by the far more dramatic and visibly active methods of conventional practice. A working woman (treated by the same Mr Wansbrough whose flamboyant military metaphors had so amused the *Lancet*) was reportedly equally dubious of her cure:

> Although perfectly freed from pain, it was enough to excite a smile to witness the woman's scepticism on the success of the operation; she could scarcely credit her senses, for when desired to turn on her back she obeyed with hesitation, and doubt, dreading lest she should encounter the 'pain'. . . It was very gratifying to see the poor creature sit up; her countenance beamed with delight, equaled only by her astonishment and grateful thanks for the 'wonderful cure' I wrought her.[26]

It is, by now, a truism that the medical encounter incorporates elements of performance. As such, it is as stylized, as subject to shifts in convention and social norms, and as deeply imbued with the prevailing *zeitgeist* as any other cultural production. At a moment when medics were fighting to wrest authority over the body away from their clients and the general public, performance was one weapon in their armoury. They certainly employed it with relish – though not without paradox. It was, after all, the twilight of 'heroic' medicine as well as the dawn of medical 'science'. The gory drama of therapeutic activities like bleeding and purging was still valued by professionals and patients alike, even while overt solicitation of public attention and approval was increasingly seen as inappropriate medical behaviour. Wansbrough presented his female patient's dubiety as a subject for pitying amusement, but her response also suggests that medical men were beginning to win their battle for exclusive rights to interpret the body. Another surgeon, operating on a local (and apparently successful) farmer, treated his cautious response with more respect. Mr Welbourn, whose bluff doubts are quoted in the epigraph to this section, remained firmly in control of his treatment. He 'allowed' Banks to use the needles, despite clearly considering the treatment useless; however, he ordered his practitioner to treat only one leg in this experiment. The needles were painful, and Welbourn compared their sensation to that produced by blistering.[27] The pains of his punctured leg were relieved, but Welbourn, unlike his doctor, was in no hurry to credit acupuncture with a cure:

> The other leg continuing to torment him, I was anxious to proceed in my work, which he would not permit – stating that he 'had, he thought, for one day had quite enough, and would like to stop to know how the left leg went on.' ... So delighted was he at having one leg and hip free from pain that, upon my departure, he wavered a good deal whether to have the other done or not; at last he determined to wait and see the result of what had been done.[28]

Only after a week of complete relief was the farmer convinced; he then promptly sought out Banks, and 'allowed me to acupuncture the other'.[29]

Case studies involving professional and wealthy patients contained fewer explicit expressions of patient disbelief – and of course, the patient's self-reports and reactions were not presented with the same amused paternalism by their practitioners. Another of Wansbrough's patients, the 'corpulent' Mr A. W., 'on learning that I possessed the means of affording him instant ease, sent for me to come to him immediately, and bring my "needles" with me'. After being cured of lumbago which had resisted bleeding, embrocations and purging, Mr A. W. 'got up ... expressing the greatest astonishment at what he termed the "magical effect of the needles"!!!'[30] A. W.'s use of the language of enchantment strikes a chord with ongoing discussions in middle- and upper-class households about another experimental cure of the day: animal magnetism, or mesmerism.[31] Indeed, the doubts which beset patients 'cured' by acupuncture suggest that lay (as well as professional) perceptions of acupuncture were influenced by the same questions about mental influence, experiential phenomena, and authority over the subjectively experienced body which would later underpin the mesmerism controversies. British responses to acupuncture foreshadow in many respects the reactions which greeted mesmerism – and the fates of these two medical innovations cast light on subtle changes particularly in professional understandings of medicine, science and cultural authority. As the speaking, self-reporting subject was being excised from science, therapies dependent on active patient participation came under threat. But while mesmerism actually relied on the patient as both an actor and an instrument for interpreting the natural world, and thus promoted an understanding of the body verifiable only through such subjective data, acupuncture explicitly relied on the patient merely by default, and presented no theory to account for the patient's experiential world. Thus therapeutic needling could remain just within the changing parameters of orthodoxy, while mesmerism unavoidably challenged those boundaries.[32]

As would subsequently prove the case for many patients cured mesmerically, A. W. directed his amazement and gratitude towards the practice of acupuncture and not the practitioner who applied it. Although in this case and several others, patients summoned a practitioner specifically to be needled – indicating that they possessed some knowledge of the treatment – they did not necessarily expect much benefit from it. For example, Thomas Martin noted that his noble patient's 'gratification was only exceeded by his astonishment'.[33] Apparently, even his daughter-in-law's eye-witness report of a cure produced by the needle had not persuaded George O'Brien that acupuncture was likely to succeed. Practitioners, like their patients, often required a comparatively higher standard of proof from acupuncture. One surgeon, William Sankey, told his audience that in each of his cases, '[i]n order to be certain of the effect of acupuncturation, no medicine or external application was employed' other than the needles themselves.[34] Given that few, if any, established treatments were displayed unaccompanied by other interventions, Sankey's strict control demonstrated both that the new 'scientific' standards were more readily applied to novel techniques, and that he believed acupuncture particularly vulnerable to challenges.

The painstaking care Sankey devoted to establishing beyond doubt the centrality of acupuncture to his patients' cures reflected an accurate assessment of the medical climate. As acupuncture was becoming visible and recognizable within the British therapeutic culture, it also became the object of less friendly scrutiny, largely from establishment physicians. The physicians, of course, had the most to lose from general acceptance of acupuncture (at least as Churchill defined it). As Vicq D'Azyr observed in the 1790s: '*toutes les maladies pour lequelles ce moyen de guérir est recommandé, sont entièrement de ressort de la Médecine*'.[35] The technique of acupuncture – the penetration of the body's surface by an implement, for a therapeutic purpose – would naturally fall into the category of surgery, and indeed, most of its early proponents in Britain were surgeons. However, many of the diseases for which Churchill and his fellow popularizers recommended the needle were ailments commonly the province of physicians – systemic illnesses like rheumatism, gout and epilepsy, rather than specific ones. In his *Treatise on Acupuncturation*, Churchill had clearly targeted surgeons, for whom the therapeutic reach of the needle was an appealing trait, offering as it did a way to expand the surgical role. This presentation of acupuncture undoubtedly restricted its audience and pool of potential supporters; only four of the practitioners reporting in *Cases* were medically rather than surgically trained.[36]

Moreover, in the *Cases*, Churchill gave a concrete example of the professional benefits that accrued to acupuncture in this respect. He described a case in which he had been called to treat a woman with a muscular injury, typically a surgically treated ailment. 'The usual remedies were persevered in' without effect until her husband 'suggested the employment of a physician' – in other words, until the failure of surgical interventions led the husband to suspect an underlying ailment requiring a systemic rather than a specific therapeutic approach. Churchill responded to this economic and social threat by offering a middle path: 'I advised acupuncturation, which was readily submitted to.'[37] The outcome, as he portrayed it, was beneficial to both patient and surgeon, curing the former, and expanding the business and the authority of the latter. These claims combined market-invasion with a break from the traditional hierarchy of British medicine. Thus it is unsurprising that it met with a fairly rapid response from physicians and medical conservatives.[38]

Even as early as 1826, the medical journals hinted at sceptical – even mildly hostile – responses to acupuncture occurring beyond the printed page. The *Lancet*, which strongly supported the technique, frequently carried acupuncture success stories taken from the records of St Thomas's Hospital.[39] In August 1826, the editor appended to such a case report a fulmination against those who disparaged acupuncture:

> We are informed that a Dr. Yeats, in the delivery of a Croonian lecture at the College of Physicians, a short time since, classed together the stethoscope, *acupuncturation*, metallic tractors and phrenology – and declared them to be alike, 'ephemeral follies'. This is, perhaps, well enough for the College of Physicians – it is quite in accordance with their *cobwebed* [sic] prejudices, but we cannot help expressing our disgust at such *bigotry*.[40]

The editor (and the comments, though unsigned, certainly have Wakley's inimitable touch) then took the battle to the College elite, implying that their own motives were none too pure: 'There is an "ephemeral folly"...which is well worthy of all satire – we mean the pompous advertisement paragraphs which...appear in the columns of a fashionable morning paper, announcing that Dr Yeats has returned...from attending Duke A. or Lord B.' As the editor noted correctly, if with somewhat unseemly glee, 'This "ephemeral folly" is intended, we have no doubt, as a *metallic tractor.*'[41] Certainly, in these early days of the *Lancet*'s history, the mere fact that the Royal College of Physicians disapproved of a technique was sufficient accolade for the *Lancet*. But the grouping of medical innovations which this passage reported, and to which the editor objected so strongly, is itself revealing. The quotation was deployed, in the first instance, to poke fun at the knee-jerk traditionalism of the Royal Colleges, and to discredit their moral and professional authority by focusing on their members' greed. However, the *Lancet* commentator was careful to ensure that his repetition of Yeats's criticism did not harm acupuncture. The association of acupuncture with the stethoscope was harmless, if not actively favourable to the needle: the generation of practitioners who were still resisting the newfangled diagnostic aid and its dangerous associations with French materialism were unlikely to incorporate acupuncture in any case, while younger doctors would find the link between acupuncture and the latest in medical technology quite appealing. Likewise, phrenology was popular, not just among the medical professionals but also with the paying public. These two links went, therefore, unremarked by the editor. However, the analogy drawn by Yeats between acupuncture and metallic tractors was a different matter. Metallic tractors were the tools by which an American doctor named Perkins had claimed to be able to cure all pain and disease (and which he had patented and sold with considerable success in both countries for several years). As a medical fad which had been recently discredited through 'scientific tests', any association between it and the already slightly *outré* technique of needling was dangerous to the survival of the latter within the orthodox medical community. Thus the anonymous editor focused his retort specifically on that part of Yeats's remarks, displacing the connection from acupuncture to advertising.

Later in the same year, the *Lancet*'s editor returned to the topic, this time in an *advisory* role. Again, the context was a case report, in which persistent sciatica was cured by acupuncture (the newly healthy patient, after two weeks of daily needling, reportedly 'has his joke, for he votes the Doctor to be "a bit of a *bore*"'). However, where the earlier article broadly attacked the motives and credibility of acupuncture's critics, in this report the editor turned his attention to describing the proper use of acupuncture. Only by implication did the article label those who dismissed acupuncture merely incompetent or over-hasty:

> We are well persuaded that many valuable remedies in particular cases have fallen into disuse, and are condemned as 'follies of the day', because they are indiscriminately used. An old friend of ours was in the habit of using three trite considerations on the subject of venesection – *the time when, the place where*, and *the manner how*. These observations are applicable to acupuncturation.[42]

Again, the use of implied analogy also served the promotional aims of the article; acupuncture was likened to a well-established technique, the mode-of-action of which was less than certain (although much theorized).[43]

From the *Lancet*'s robust rebuttals, it is apparent that opponents of acupuncture typically described it either as a fad or as therapeutically weak, rather than as quackery. The *Lancet* actively condemned such portrayals as 'the sneers of certain learned sages'.[44] Indeed, the level of the *Lancet*'s activity is one of the few signs that the technique was so regarded – the Royal Colleges made no formal statements or policy against the use of acupuncture, and certainly admitted acupuncture users to membership.[45] However, while physicians continued to assert their primacy over revenue-producing illnesses like rheumatism and neuralgia, surgeons were exploring new and more profitable avenues of expansion than could be opened by acupuncture.[46] Unlike surgeons, physicians had no traditional preference for external applications. Indeed, their initial hostile response to the use of the stethoscope in diagnosis reflects their resistance to even such circumstantial linkages with the surgeons (whom they regarded as a lower order) as might be fostered by the visible use of instruments and the hands.[47] Thus acupuncture had little immediate appeal to physicians, especially as its British proponents frequently asserted the importance of anatomical knowledge for its users. Those physicians who did support acupuncture, like John Elliotson at St Thomas's Hospital, tended to use it alongside more conventionally medical remedies. By the early 1830s, when the first flash of interest in acupuncture had well and truly faded, even Elliotson was recommending internal treatments like colchicum or iron for severe cases of rheumatism and gout (although he did simultaneously mention acupuncture's usefulness in mild chronic cases). Driven to respond, Churchill more strictly limited the circumstances in which he recommended acupuncture, but did not eliminate 'medical' ailments from his list of the needle's appropriate targets.

Acupuncture, empiricism and scepticism

> There are some modern practitioners, who declaim against medical theory in general, not considering that to think is to theorize;...and happy therefore is the patient, whose physician possesses the best theory.[48]
>
> Erasmus Darwin, 1794

Churchill's second monograph on acupuncture, with its 'appendix' of cases, appeared in Britain just as acupuncture's prominence in the medical press began to subside. Acupuncture had produced six years of mixed results since his *A Treatise on Acupuncturation* came out in 1822, and measured tones were replacing the initial enthusiasm expressed, for example, by the *Lancet*. But although Churchill's new book was indelibly marked by shifts in the decade's response to the technique, he did not greatly retrench on his early claims for the needle. He intended the *Cases* rather to solidify acupuncture's status in the canon of British practice, and to prevent further slippage in its popularity. Churchill declared this aim openly: 'I shall, in the following pages, endeavor to substantiate the claim which acupuncture so deservedly has to the attention of those, who still remain sceptical as to its effects.'[49]

Churchill claimed in his introduction that the prominent practitioner, Dr Matthew Baille, had convinced him of 'the necessity of publishing additional cases of the success of this therapeutical agent'.[50] Not unreasonably, Churchill attributed some of the continued scepticism to the fact that 'novelties in the curative art' generally met with extreme responses, with people either espousing them wholly and rashly, or ignoring them completely:

> Thus it has been with the subject under consideration [acupuncture], for while many have never practised it, others have expected too much from it, and after a few indiscriminate trials, have abandoned it as useless. But I am happy to produce confirmation of its magical powers, from the pens of men, whose veracity and disinterestedness cannot be doubted.[51]

Evidently, Churchill was concerned primarily with assuaging the doubts of his fellow-professionals, indicating the growing authority of the medical profession and their correspondingly increased control of the therapeutic encounter. While Churchill clearly remained convinced of acupuncture's value – 'its magical powers' – the tone of this introduction contrasted sharply with his earlier writings, including the exultant 1823 article in which he announced this 'appendix' as forthcoming.

In 1823, Churchill had expected scepticism to greet his monograph on acupuncture: 'When I published my little treatise on acupuncturation, I expected to be questioned about it by individuals...too polite to tell me that I had asserted what was not true; at the same time that their countenances clearly indicated the incredulity with which they viewed it.'[52] He had detailed his concern that acupuncture be used appropriately 'because many valuable remedies are lost sight of, from being injudiciously employed by those who are too fond of *analogical deductions*'.[53] However, writing a year after the publication of his *Treatise on Acupuncturation*, Churchill also clearly felt that acupuncture was, if not completely established as a part of orthodox surgical practice, at least well on its way to general use and acceptability:

> Its success has now been so conspicuous, that I can assume an air of triumph, and dare anyone to express his disbelief in what I have asserted respecting it. I am continually hearing of successful cases from respectable members of the profession...I select the subjoined cases for the perusal of your readers, that they may be induced to practise an operation that is so simple, so painless, and so convincingly efficacious...[54]

However, when the *Appendix* finally appeared, it was framed as a defensive action rather than a proclamation of acupuncture's adoption. The medical journals contained few, if any, criticisms of acupuncture's individual supporters. In comparison with their sarcastic treatment of the medical men who promoted animal magnetism or homoeopathy, acupuncturists were left unscathed. The harshest comment I have found in respect to Churchill was a criticism of his 1823 exultation as 'the language of youth'.[55] However, Churchill himself claimed to have been personally attacked: 'For the part I took in advancing the practice I have been assailed by some with unmerited abuse, while others have pitied me as a visionary, and considered the relief ascribed to it, to be the result of mental influence over the corporeal sufferings of those, whose understandings are weak.'[56] The reference to 'mental influence' was a thinly veiled allusion to animal magnetism, a practice rapidly becoming both visible and unacceptable in orthodox medical circles. Churchill responded to it with the same sharpness shown by the *Lancet*'s editor when acupuncture was compared to metallic tractors. Although Churchill did not openly accuse these unnamed attackers of medical bigotry, he suggested it. At the same time, he implied that they were irresponsibly and unscientifically over-simplifying the relationship between the mind and the body: 'the latter ought to be reminded, that their explanation of its [acupuncture's] success, involves in it a subject of physiological inquiry, quite as intricate as any other causes for the unintelligible phenomena, which accompany the actions of needles when inserted into the various tissues of the body'. Notwithstanding his own belief in the therapeutic power of imagination, Churchill strategically disputed the assertion that acupuncture acted by mental influence as requiring 'a mode of reasoning that it would be very unphilosophical to indulge in, and which is at utter variance with the general laws of nature'.[57]

In building his bulwark around acupuncture's gain, Churchill chose his materials and evidence quite carefully. For example, he used the French physician-acupuncturists, and their experimental explorations of the properties of acupuncture as evidence supporting the use of the therapy: 'The French, with their characteristic zeal for the advancement of medical science, practise this neglected operation with increasing success; and the results of their investigations . . . verifies the praises which have been bestowed by others upon it.' However, Churchill also used *Cases* to distance himself from the French experimentalists – again, a strategic move designed to sever ties between acupuncture and the excesses of French medical and political radicalism on the one hand and their despised Gallic 'effervescence' on the other. Although he repeatedly expressed his interest in experimental studies of acupuncture, Churchill observed that the French experiments had produced no useful information.[58] He presented their work as, in fact, having failed to discover an adequate rationale for the cures produced by acupuncture and tarred French practice with the traditional English accusation of flightiness: 'many more suppositions, and fanciful ideas, have been indulged by others of the French; and they have been led to practice acupuncture in cases particularly unadapted for it'.[59] In the end, experimentation was worthy of notice only when it was subjugated to empirical aims. These somewhat contradictory statements paint an amusing picture of Churchill weighing the authority attached to French clinical and experimental science against the instinctive distaste for things French among his colleagues. Clearly, Churchill knew that experimentation, with its implied rigour and record of attacking medical fraud, would be a sturdy support for acupuncture, if convincingly deployed. However, he was equally aware that an association with France – to say nothing of more exotic nations – could easily undermine the technique's credibility in Britain.

The scientific investigations were, Churchill concluded, still valuable as models for future work. Indeed, he finally acknowledged (as he had not done in either the *Treatise* or his shorter essays) that theory could play a role in shaping acupuncture practice: '[C]ould a rational theory be established, its practical utility would become much more efficiently manifested, by the precision with which we could adapt it to individual cases of disease.'[60] For the first time, Churchill included copious descriptions of the various theories and their flaws; he also justified his own failure to perform 'such experiments as the subject demands' with the excuse that his personal practice was too small to support it. Without coming out explicitly in favour of any theory, he presented one as the most likely:

> It has long been supposed that the nerves are the *media* by which a fluid [a word which he qualifies as 'in absence of one more definite, and in accordance with popular opinion'] analogous to the galvanic, is circulated or conveyed to the remotest parts of our structure;... and as the effects of acupuncture are so instantaneous, it is very natural to infer, that they proceed from, or are effected by, some principle, like to, or connected with, the electric, pervading the animal machine.[61]

His hypothesis was firmly based on the latest theories of the nervous system, but Churchill offered only his own practical experience of acupuncture, and the experiential evidence of his patients' senses to support it. He described his patients' self-reported muscular twitchings, numbness and – intriguingly – that 'the patient frequently experiences sensations at a remote distance from it, resembling an electrical aura'.[62] Churchill noted that the feelings he described were the same as those reported by Pelletan and Cloquet, and that they discovered and described these feelings completely separately. He argued that 'the effects which we felt, must [therefore] be ascribed to a real cause for them, and not to mere chimerical ideas'.[63] His statement also gives the modern reader some evidence on which to judge his own technical skills; the sensations he described were, as in his first essay, identical to those described in the Chinese and Japanese traditions as resulting from the correct penetration of an acupuncture point.[64]

Churchill explicitly reaffirmed his conviction that acupuncture did produce actual 'physiological changes'. However, he still saw it as worthwhile to deny any firm and exclusive attachment to a particular theory, including the one he advanced himself.[65] Instead he aimed to counter the blurring of technical definition which he sensed was reducing the effectiveness and lowering the credibility of acupuncture. Perceptively (and consistently), he judged that the truly appealing aspect of acupuncture was its success rather than its fit with one theory or another, and he framed his argument, finally, as atheoretical:

> [Acupuncture] is, at present, a mere matter-of-fact business; and our ignorance is the less to be regretted, while I can state, from personal observations of some years, and from information derived from others, that in those diseases for which I have particularly recommended it, it often effects a cure after all other apparent means have failed: sometimes immediately; at other times after several days' repetition.[66]

Unsurprisingly, he continued to tout the safety of acupuncture – comparing its few bad effects to those produced by bleeding:

> I have never known an accident to arise from acupuncture, nor a single untoward symptom to have been produced, unless a slight degree of faintness should be so considered: but even this may be ascribed, with great propriety, to the operations of the mind; as I have known it to occur merely at the thought of the operation, as well as of many others, amongst which bleeding may be familiarly ranked. [Note again the analogy to the mysterious but firmly rooted remedy of bloodletting] [67]

Churchill then offered his carefully selected cases; their textual mass and the vivid manifestations of acupuncture's healing powers which they detailed illustrate his continued reliance on empirical rather than theoretical criteria of success and suasion. The cases which Churchill selected specifically portrayed empirical successes swaying both patients and practitioners; of course it is impossible to know whether his selection incidentally or deliberately justified his strategy.

1829–40: subsidence

Churchill's *Cases* did not receive the generous press coverage given to his *Treatise on Acupuncturation*; in fact, I have found no mention of it in the medical press. This lacuna was symptomatic of a broader lack of journalistic interest in acupuncture as a means by which to cure or relieve pain; from 1820 until 1830, the *Lancet*, the *Medical and Physical Journal* and the *Edinburgh Medical and Surgical Journal* had between them published twenty-nine articles on acupuncture. Between 1830 and 1840, the same three journals published twenty-one articles on acupuncture. Although numerically the journals' second decades much resemble the preceding one, they in fact reflect a very different level of response to the use of the needle against rheumatic and neuralgic pain and disease. By 1830, another medical use of the needle was sharing the name 'acupuncture' with the British adaptation of the Asian therapy; the second technique involved repeatedly piercing the skin with a needle to release oedematous fluid. In the first decade of British acupuncture, the dominant meaning of the term was that referring to the relief of pain – the therapy which had grown from originally Chinese roots. Only two of the twenty-nine journal articles described the use of the needle as a safer lancet or trochar as 'acupuncture'. However, after 1830, the two meanings were co-dominant and often conflated – Churchill's singular needle had not survived the decade of its conception. Of the twenty-one articles discussing 'acupuncture', nine used the term to describe the mechanical alleviation of oedematous conditions.[68] This often successful operation was solidly grounded in the surgical tradition; in treating the oedema symptomatic of several constitutional disorders, these 'acupuncturists' made no claims about the underlying medical conditions. Thus, although the use of the term was contested by some observers, many of acupuncture's supporters (notably Churchill, by 1828) accepted the secondary meaning and the cures recorded under it. The curative needle was threatened with disappearance, and its more pragmatic proponents sought goodwill and publicity wherever it could be found.

It is difficult to reconstruct the nature and effects of historical absences – lacunae can be at best suggestive on the subjects of causation and perception. But of course, the long public decline of acupuncture was naturally characterized by the silence of press, public and profession on the subject of therapeutic needling. Fortunately, the handful of articles on acupuncture which appeared in the medical press between 1828 and 1856 do cast some light on to the reasons for their own rarity.

John Renton, a Scottish physician, presented several successful acupuncture case studies in 1830. He acknowledged that, 'it does not appear, if we may judge from the few cases upon record, that the practice [of acupuncture] has been generally adopted in this country'.[69] After strongly supporting the technique, Renton offered his own explanation for acupuncture's failure to thrive:

> if the system is too much undervalued now, it is equally true, that it was very much overvalued by those who first recommended it to notice. The utility of a specific is very readily suspected when its infallibility is given out for the removal of too many diseases, and more particularly of those between which no analogy can be traced...[70]

But mere disappointed expectations, although damaging to any novel medical practice, were not the sole, or even the most important factor in British resistance to acupuncture. Rather, Renton regretfully acknowledged that any cure with a mysterious mechanism would be naturally suspect:

> when...no satisfactory explanation can be afforded of the modus operandi of the reagent, professional people, unhappily for the interests of medical science, are too apt to reason upon the authenticity of the facts averred, instead of adopting the more simple and direct method of determining their value by subjecting them to the test of further experience.[71]

In other words, therapeutic claims pressed on the basis of empirics rather than theory were inevitably judged doubtful or meretricious.

For Renton, himself active in politics, the impact of theory on professional responses to medical innovations would have been especially clear; this was a period of particular turbulence within the medical professions, when the links between politics and medicine were more visible and more hotly debated than ever before.[72] From the power of mesmeric trances to physical examination to new interpretations of comparative anatomy, the theories by which medical novelties were explained and authorized were inevitably politically inflected. These inflections shaped the reception given to each innovation by the different factions of the medical community.[73] Churchill probably intended his theory-free presentation of acupuncture to render it available to the broadest possible spectrum of users. However, his caution may have reduced acupuncture's appeal in a period when the medical periodicals were themselves active participants in the debates of the day. With the medical press divided into opposing camps, an innovation's ability to strengthen or weaken one side of the politically imbued debates was as important in determining its access to print as its healing potential.

Despite acknowledging that the enigma of acupuncture's *modus operandi* hampered its acceptance and impugned the credibility of its supporters, Renton did not place a great premium on discovering the mode of action. In fact, he argued that the attempts to account for its effects had themselves harmed the popularity of the therapy: 'Indeed, the different attempts which have been made to account (as by electricity for example) for the various physical and physiological phenomena produced by acupuncturation, have been very injurious to the successful diffusion of the practice.'[74] Renton, unfortunately, did not specify how the experiments had harmed acupuncture's image, but presumably it was by persistently failing to produce convincing evidence to support any explanatory theory. With some frustration, he described the (perhaps inevitable) result of Churchill's empirical strategy in a medical culture which was increasingly interested in scientific models of causation and proof:

> [A]ccordingly we find, that the very rapidity and perfection of the cures have acted as causes why the efficacy of the remedy has been doubted, and that its boasted remedies have been imputed more to mental action...than to any real good effects resulting from the operation itself.[75]

Many of the articles on pain-relief acupuncture which did appear after 1830 had foreign origins, indicating that European medics continued to use acupuncture, and that the journals were at least still willing to publish on the subject. Clearly needling had not suddenly been rendered discreditable by some event which escaped the written record.[76] Doubtless the authority of European reports contributed in some degree to the maintenance of low levels of acupuncture use in Britain. However, given the periodic bouts of xenophobia expressed particularly within orthodox English medicine, these foreign sources would not have been as persuasive as domestic accounts. For this deficiency, Tatam Banks's 1831 article suggested another explanatory factor, and one which may have proven crucial in reducing acupuncture's visibility in the medical press. Acupuncture was increasingly perceived to be impotent in cases of acute muscular or nervous pain. Clearly, these were exactly the cases in which the patient would have been most likely to call in a surgeon, and despite the example of Scottish physicians like Renton and Banks himself, surgeons were still more commonly acupuncture-users than their medical colleagues.

This surgical predominance, rooted in the traditional designation of external and internal treatments as respectively surgical and medical, and sponsored by Churchill's initial presentation of acupuncture, greatly complicated the process of educating the next generation of potential acupuncture-users. At this crucial point in the extra-textual diffusion process, the transmission of knowledge about acupuncture seems to have stalled. The scepticism which surfaced in Churchill's accounts of the happily healed, and which pervaded Dendy's re-telling of the Earl of Egremont's cure, took far less benign forms in the lecture theatres and hospital wards where acupuncture was presented to the crucial next generation of practitioners. As acupuncture's first British decade closed, those students who were exposed to the technique heard it described in cautious tones and understatements, if they heard of it at all. An 1832 clinical lecture at St Thomas's Hospital, presented by Dr John Elliotson, exemplifies the climate of doubt, and sheds some light on its causes. Elliotson was describing a case of rheumatism which he had treated that week at the hospital. He had employed acupuncture, and the patient had left the hospital cured. Yet Elliotson's review of the case showed no vestiges of the triumphalism which had marked even his own earlier reports of the healing needle. Instead his tone was disbelieving: 'The case of rheumatism I will not pretend to say was real, and, therefore, I cannot say that I cured the patient; but he went out well.'[77] The doctor's uncertainty was directed in the first instance towards the patient's self-report of his pain – 'He said he had a violent pain in the back, that was relieved by heat, but *I could not see the pain or its effects*. It was not attended by quickness of the pulse, or foulness of the tongue.'[78] The pain was not physically marked upon the body, not externally legible, and was thus automatically suspect. Elliotson's medical response thus had two goals; his first, and ostensible aim was to cure the patient; his second, made explicit only to his medical audience, was to establish an independent, external assessment of the patient's complaint. In practice, Elliotson reconciled these goals by choosing acupuncture as the therapeutic instrument. Thus, the needle was employed to test the patient's veracity, serving either to cure or to punish him.

> He said he had pain, and, as I have just observed, relieved by heat, and therefore it was a proper case for sticking needles in his back. *If he was shamming, I knew he would not like the remedy; but [if] it was real, acupuncture was a proper measure*. He had a needle introduced on each side of the back, which was allowed to remain two hours every day, and in three or four days he said he was perfectly well. [79]

Yet although it was the patient whom Elliotson expressly doubted, acupuncture too came across as suspect. The dubiety surrounding the very existence of this man's 'rheumatism' certainly lessened the impact of an acupuncture cure for it. The selection of the needles as suitable punishment, if the patient was in fact 'shamming', cast them in no pleasant light either; acupuncture, in this scenario, was at best harmless but intimidating. At worst, it was sufficiently terrifying to scare off a man hardened to deceit. Finally, even considering the case in the most favourable light, acupuncture was presented as curing only such trivial pains as could be 'relieved by heat' and which was unaccompanied by physical disease – hardly a daunting or unique brief with which to justify the use of a novel and mysterious medical technique.

Elliotson was not speaking solely to the students at St Thomas's Hospital who surrounded him. Rather, at this point the professional paths of diffusion reunited: Elliotson's lectures were syndicated by the *Lancet*, and thus were addressed also to the far greater, if less privileged, crowd of practitioners and students who read his lectures there, week by week. Later, they were collected into several highly orthodox multi-edition collections of his *oeuvre*. It is worth noting that in this censored form they retained their references to acupuncture but were scrupulously edited to remove all mentions of the mesmerism which led to Elliotson's medical fall from grace. Elliotson had been an early and prominent supporter of acupuncture, and it is certainly possible that this very public demonstration of his own continued (if not convinced) use of needling was intended to buoy the technique. But acupuncture's support was suffering largely because of a general trend in patient care – and specifically from a change in the form of the medical encounter. Western doctors were finally overtaking traditional Chinese physic in the process of creating an externally legible body and in establishing the medical man as the authoritative reader of that physical text. The patient's experiential pain and symptoms, if not inscribed on the body in a way accessible to external examiners, were – at least ideally – no longer central to diagnosis.

In Churchill's arguments, acupuncture was defined almost exclusively by its analgesic efficacy; it could cure or relieve patients when they had no organic illness (as signified by the very visible signs of inflammation or fever), or when they had 'nervous affections', a category the vagueness of which hints at the diagnostic invisibility of its constituent ailments. But these complaints, if they were to be taken seriously, required a level of patient involvement which no longer fit the profession's (aspirational) paradigm of the medical encounter. The level of control exercised even by poor patients in reporting both their symptoms and the degree to which they were relieved by acupuncture was far too great to be readily accepted by a medical profession which was building an exclusive authority over the experience of illness. Elliotson's new-found scepticism exemplifies this shift.

1840–70: submergence

> The Empire of Medicine has just passed through one of those unaccountable paroxysms of credulity to which, from time to time, it seems ever to have been subject...folly will have its turn. Even Medicine is not always vigilant and sometimes Aesculapius nods.[80]
>
> A Surgeon, 1839

Where the medical tone of the 1830s had been turbulently radical, the decades which followed were defined by the medical establishment's minor concessions to the moderate reformers and uncompromising rejection of the radicals.[81] In this climate, and with the Opium Wars further tainting British responses to China, the moderate reformers who were the backbone of British acupuncture use had little to gain by publishing additional case studies on acupuncture. The radicals who might still have benefited from – and would certainly have enjoyed – acupuncture's establishment in the face of Royal College mockery had limited access to the medical periodicals and to the 'clinical material' necessary to the production of acceptable articles. Consequently, during the 1840s and 1850s, the medical periodicals were almost completely silent on the subject of acupuncture. The *Lancet* and the *Edinburgh Medical and Surgical Journal* each published only one article on the therapeutic needle between 1840 and 1849. Both reported on purely mechanical uses of acupuncture, but the Scottish article, an extract from the German *Journal für Chirurgie und Augenheilkunde*, left open the possibility of a slightly more complex *primum mobile*. The incident recounted (in the *Journal*'s column for 'Surgical Pathology and Therapeutics') involved the reduction of a strangulated hernia, usually accomplished by a complicated and dangerous operation:

> Dr Daser, before having recourse to the operation for strangulated hernia, with a view of trying the effect of acupuncture, for the purpose of evacuating the gaseous contents of the strangulated portion of the intestine, made two punctures in it with a long fine needle. No

gaseous matter apparently escaped, but the patient complained of acute pain, and loud gurgling sounds were heard in the abdomen, immediately after which the hernia was spontaneously reduced. Dr Daser attributed this fortunate occurrence to the prick of the needle having excited the contractility of the intestine...[82]

The *Provincial Medical and Surgical Journal*, precursor to the *British Medical Journal*, printed two very different articles referring to acupuncture. The first was an abstract of Italian experiments on acupuncture's efficacy in cases of asphyxia and drowning. Such foreign reports typified the vehicles through which information about acupuncture was disseminated in the medical media in this interval of submergence. However, the second article was the text of an after-dinner speech given to the Westminster Medical Society in 1842; its subject was Chinese medicine, and it briefly mentioned acupuncture. This essay was an exemplar of a new and rapidly growing genre of medical writing, directed towards the satisfaction of curiosity (and measured self-congratulation) rather than the interrogation of new medical techniques. Within that genre, the article was unusual only in being so fully reported by an established medical periodical; this anomaly can be explained by the fact that the paper was read to an established medical society, whose meetings were regularly reported in the *Provincial Medical and Surgical Journal*. Usually, articles discussing Chinese medicine appeared in the burgeoning sector of Asia- or mission-focused (and produced) journals, like the *Chinese Repository*, printed in Canton from 1832. The *Chinese Repository* was available in Britain, and was popular enough to go through two editions as annual volumes. By the 1840s, it was publishing translations and commentaries on classical Chinese medical texts, as well as their philosophical and religious counterparts, and general gossip about the British expatriate community in China.[83]

While the number of articles specifically discussing pain-relief acupuncture declined, coverage of China and Chinese medicine in general increased, particularly during and just after the Opium Wars, when interest in and access to China was great (if predominantly hostile). Once again, missionary activity was fostering cross-cultural transmission of knowledge. This time, however, the missionaries were British, Protestants and, if medically trained, trained also to expect little practical benefit from the study of non-western medicine. Furthermore, after the 1820s, they were no longer providing information about an unknown phenomenon when they described acupuncture. Despite the ambiguities surrounding the therapeutic needle in European and British practice, it had acquired a western pedigree. Supporters and detractors of acupuncture in Britain turned to the European literature for evidence and interpretive authority, rather than to missionary accounts. This choice is unsurprising, as views of Chinese medicine presented by observers in China heavily emphasized the exotic and bizarre; they were rarely flattering or even necessarily recognizable portraits of the indigenous medical practices.[84] Thus, one doctor published a diatribe against Chinese medicine in which much was made of the exotic names given to variants of the pulse, and the more alien aspects of Five Element theory were lovingly detailed. He then described acupuncture in China: 'where a bone or muscle or joint is in a state of inflammation, chronic or acute, a substantial stiletto is thrust into it and stirred about in a most reckless manner producing the most terrible consequences, sometimes causing death, or making the patient a cripple for life'.[85] Moreover, the medical missionaries who provided British authors with their new material were as materially interested in downplaying Chinese medical expertise as their Catholic predecessors had been in polishing China's image. The medical missions had, after all, been founded and funded on the twin assumptions that China was medically ill-served and that the provision of superior western medical care would open the way for Christianity and civilization (and of course, the profitable trade which seemed inevitably to follow). Obviously, a picture of Chinese medicine as not merely occasionally effective but as worthy of adoption, or of Chinese medics as competent healers would conflict with the narratives of suffering and squalor which elicited mission funding and justified their somewhat controversial medical focus.[86]

The visible needle: sites of persistence, 1840–70

Unlike the periodical press, medical compendia and dictionaries frequently discussed acupuncture in greater detail – although not necessarily with greater approbation – as the century progressed. This shift began as new editions were released in the late 1830s and 1840s; the additions and expansions reflected a continuing exploration of acupuncture which although limited in its nature and aims, far exceeded that indicated by the periodical coverage. In 1820, Robert Hooper's *Lexicon Medicum* described acupuncture as 'bleeding performed by making many small punctures'.[87] In 1839, the seventh edition ('revised, corrected and enlarged' by a physician named Klein) included a more substantial entry, which began by correcting the idea that needling was a form of phlebotomy. Acupuncture was redefined as 'an operation which consists in the introduction of a needle into any part of the body with a view to the relief or cure of disease'.[88] The entry referred to acupuncture's Asian origins but expended more effort on a description of its passage through France and Europe, noting experiments on animals and the modification of acupuncture to combine it with galvanism. Klein made it very clear that the British model of acupuncture practice differed from (and by implication, was superior to) its European and its Asian usages. He first observed that 'British practitioners confine this operation to muscular, tendinous, and aponeurotic [nervous] parts; but the Orientals pierce the abdomen with needles for relief of colic and other affections.'[89] Later, he expanded, without enthusiasm, on the variety of illnesses for which acupuncture was employed in China and Europe:

> Acupuncture is employed in a great variety of diseases. In China and Japan it is used in abdominal affections, apoplexy, convulsive diseases of all kinds, fever, gout, rheumatism, gonorrhoea, and many other cases. On the continent of Europe it has been tried with more or less apparent success in most diseases of which pain is the principal symptom – as rheumatism, neuralgia, gastrodynia, pleurodynia, headach, toothach, &c.[90]

Once again, he followed this information about what he clearly regarded as the irrational enthusiasm of the Asians and Europeans for the needle with a contrasting description of the sensible and limited practices of the British Isles:

> In this country, acupuncture is not generally considered worthy of any confidence, except in cases of local pain, quite unattended with any inflammatory action: that form of chronic rheumatism in which the nerves are chiefly implicated...is the disease in which this remedy has been found most unequivocally useful.[91]

Finally, Klein noted that the use of the needle 'to afford exit to the fluid effused in anascara and oedema' was both more effective and 'much less dangerous' than the traditional treatment of scarification.[92] Although decidedly lukewarm in his praise of acupuncture, Klein's entry included all of the information about the operation of acupuncture which was considered requisite to its practice in Britain. The novice practitioner could learn from this text in what conditions to use the needle, how to site its insertion, how exactly to insert it through the cutis, and how long to leave it embedded in the flesh.

John Elliotson's collected lectures, published in 1839 as *The Principles and Practice of Medicine*, gave its readers even more advice on treating patients with acupuncture. These volumes described acupuncture in the context of the single illness for which Elliotson still considered needling appropriate: chronic or acute rheumatism without inflammation. Although Elliotson was dismissive of acupuncture's utility in other complaints, for this form of rheumatic pain, he remained an enthusiastic supporter of the needle: 'It is in this description of the complaint that you will find acupuncture of great use...Acupuncture is not an absurd remedy. It is a strong one; but [sic] I am quite satisfied that it is a real remedy, if it be properly applied.'[93] His depiction of acupuncture in the main text included no information about its origins; only the technique and uses of needling were detailed. However, the *Principles* included an appendix on acupuncture drawn from his contribution to the *Cyclopedia of Practical Medicine*. Here, acupuncture was portrayed as having wide-ranging applications and benefits, and the reader was reminded at length of the needle's exotic history, Asian and European.

Medical dictionaries, though naturally less detailed than the medical compendia, also slightly expanded their coverage of acupuncture after 1830. Because their entries were so compressed, the information which was selected for inclusion reveals what their editors considered to be essential to acupuncture practice (and appealing to their readers). An 1845 dictionary, for example, offered its prospective audience of medical students three facts with which to sum up acupuncture: 'Acupuncture:...an operation originally practised in China and Japan. It consists in the adroit *introduction* of a gold or silver *needle* into various parts of the body affected with pain or swelling; and is sometimes productive of temporary relief. The modus operandi of the remedy is not obvious.'[94] Acupuncture's connection with Asia and its mysterious mode of action were still worthy of note, as was the unpredictability of its therapeutic effects. As with other minor operations defined in its volumes, the *Pentaglot Dictionary* then referred its reader to a more detailed source – in this case, Churchill's *Treatise on Acupuncturation*.

Robley Dunglison, educated in Edinburgh and London, a Fellow of the Royal College of Surgery, and a prominent promoter of moxibustion in the 1820s, initially approved of acupuncture. However, he gave an ambivalent description of needling in his 1842 edition of the *Medical Lexicon: a New Dictionary of Medical Science*.[95] Dunglison first explicitly categorized acupuncture as 'a Surgical operation'; then revealed that it was 'much in use amongst the Chinese and Japanese'; and finally that it consisted of 'puncturing the parts with a very fine needle'. He noted its recent popularity for 'obstinate rheumatic affections, &c. and apparently with success'. [96] Then, somewhat bizarrely, Dunglison observed that acupuncture was also a mode of abortion 'in some countries' – hardly a fact likely to recommend the treatment to the respectable medical practitioner, especially in an era when the profession was struggling to establish its moral authority.[97]

In his 1843 volume *New Remedies: Pharmaceutically and Therapeutically Considered*, Dunglison described acupuncture in greater detail but with much the same emphases. He observed that, 'Although acupuncturation is really an ancient therapeutical agent, attention to it has been so much revived of late years, and its use has been so largely extended, that it may be looked upon as constituting one of the novelties of therapeutics.'[98] Again, acupuncture was described in terms of inserting needles into the body to relieve pain. It was further estranged from European traditions through Dunglison's comment that it was unknown to Greeks, Romans and Arabs, and indeed had been used only by the Asians. Dunglison credited the experimentalist and clinician Jules Cloquet with the revival of acupuncture in France and Europe generally, and painted a vivid if apocryphal tableau of its use in the hospitals of Paris:

> [Acupuncture] was for a long period a fashionable article in the hospitals; so much so, it is affirmed, that attempts were even made to heal a fractured bone by it without the application of any appropriate apparatus! and at one time, it is said, the patients in one of the hospitals actually revolted against the *piqueurs médecins*.[99]

His tone in this passage indicates the degree to which French authority had been devalued in British medicine. Dunglison then moved to the mechanics of puncturing, indicating that puncture was always *in loco dolenti*. The dependence of the physician on his patient in this stage of the operation was considered worthy of remark, but was not absolute: 'where the feelings of the patient do not point out the spot, it must be suggested by our knowledge of anatomy and physiology'.[100] Finally, Dunglison turned to the *modus operandi* of acupuncture. Unlike his predecessors, he asserted that acupuncture undoubtedly operated through 'a new nervous impression, produced by the needle in the parts which it penetrates'.[101] Dunglison also left his readers in no doubt that he considered this to be a very weak source of power over the body. Thirty years later and teaching in Philadelphia, Dunglison completely repudiated acupuncture, explicitly because of its 'contemptible' Chinese and Japanese origins.[102]

Acupuncture's origin was considered a weak point by many medical commentators writing during and after the Opium Wars. While even its less convinced supporters drew careful lines between acupuncture practice in Great Britain and in Asia, its critics took the opposite tack. In 1864, Robert Kemp Philip, writing for the educated lay market, described acupuncture as 'a very painful mode of curing certain chronic diseases by puncturing the part freely by one or a series of sharp, strong needles'.[103] He then closely linked its British practice with the mode of needling used in China. Unlike advocates of the needle, Philip claimed that '*The only difference* in the practice consisted in the Eastern surgeons using pure gold for their needles, while the Europeans employed those made of steel.'[104] He dismissed its early popularity, ascribing it to the faddishness of medicine: 'like all new fashions in medicine, [acupuncture] was, for a time, extensively adopted', and concluded optimistically: 'The practice is now almost abolished from English surgery.'[105]

In fact, and doubtless to Philip's disgust, acupuncture was well represented even in the medical compendia designed for the household market. In *Beeton's Medical Dictionary*, acupuncture was give a short but fairly complete citation, mentioning 'the East', but focused on European and British practices and innovations:

> It has been practised both in Paris and England with satisfactory results in different kinds of diseases, primarily neuralgic pains and chronic rheumatism. The needle, which is of steel, is passed by a slight rotatory motion to the required depth, and allowed to remain from a few minutes to several hours. The needles are sometimes used as conductors of the galvanic fluid to the deep-seated parts, and are sometimes made hollow in order to convey some sedative solution.[106]

Haydn's Dictionary of Popular Medicine and Hygiene was similarly brusque in its acknowledgement of the Chinese. Its editor used the space instead to propose a mechanism for acupuncture's activity: 'It has been suggested that the relief caused by the proceeding is owed to the fluid contained in the nerve sheaths being thus allowed to escape. It is a very favourite proceeding with the Chinese.'[107] All of the popular sources, whether advocating or denigrating acupuncture, agreed on two aspects of acupuncture: its goal was the relief of chronic pain and diseases characterized by pain, and its mode of application was that of 'thrusting needles...into the painful part'.[108]

Intriguingly, durable publications which specifically targeted surgeons did not always follow the trend of extending their coverage of acupuncture. *The Surgeon's Vade Mecum*, published in 1839, grouped acupuncture with the minor operations. Its editor, Robert Druitt, summed up the treatment in two dismissive sentences: 'Acupuncture is easily performed by running in a sufficient number of needles with a rotary motion. Its utility is very problematical.'[109] In 1842, another surgical compendium mentioned only the use of the needle to treat oedema – and that technique was not even called acupuncture.[110] A textbook on minor surgery published in 1866 did include acupuncture for the relief of pain, but only in a short section describing counter-irritants: 'Acu-puncture...is generally classed among the counter-irritants. It consists in introducing needles into the tissues, and allowing them to remain there for a certain time. Acu-puncture is most frequently employed in painful nervous affections, especially in sciatica.'[111] Rheumatism, lumbago and gout were not among the many conditions which this textbook taught young surgeons to treat, and acupuncture was not mentioned in the treatment strategies listed under individual ailments even for tetanus, oedema or sciatica. In 1884, *The Science and Art of Surgery. A Treatise on Surgical Injuries, Diseases and Operations* confined its definition of acupuncture to the operation of puncturing to reduce oedema.[112] None of these surgical texts acknowledged the Asian origins of acupuncture – a lapse excused perhaps by the fact that the 'acupuncture' they described was a western creation, albeit one triggered by the availability of knowledge about Chinese forms of needling.

The Science and Art of Surgery had been revised and edited by an assistant professor of Clinical Surgery at University College Hospital named Marcus Beck, who two years previously had been involved with the compilation of a state-of-the-art *medical* dictionary. This volume, *A Dictionary of Medicine, Including General Pathology, General Therapeutics, Hygiene, and the Diseases Peculiar to Women and Children*, contained a long and generally positive description of acupuncture, written by Beck. In his entry for acupuncture, Beck described acupuncture as 'an ancient mode of treatment for the relief of painful affections, now but little used, consisting in the introduction of fine round needles through the skin to a varying depth'.[113] He noted that knowledge of the technique had been transmitted to Britain from Japan and China two centuries previously, and that the technique was used in Britain only for lumbago and sciatica, 'in which affections it undoubtedly gives relief'.[114] Beck subsequently turned to a careful, step-by-step description of the entire operation – a description which clearly reflected personal experience with the technique:

> The operation is thus performed. The patient is laid upon his face, tender spots are sought for... The needles are then pushed in vertically for a depth of from one and a half to two inches, and allowed to remain from half an hour to two hours... In sciatica it is recommended, if possible, to make the needle actually penetrate the nerve. This is known by the patient complaining of a sudden shooting pain down the back of the leg.[115]

The success of the operation obviously still relied heavily on the patient's experience and self-reports, despite Beck's careful use of physician-centred terms. Beck had his own opinion about acupuncture's mode of action, at least in some cases. The theory he supported combined the materialist interpretations of acupuncture's early history in Europe with newer anatomical and physiological knowledge: 'The mode of action is uncertain, but in sciatica, it has been supposed that the puncture of the nerve sheath allows the escape of fluid.'[116]

Beck's article also indirectly addresses the reluctance of regular medical journals to print articles on acupuncture. In the course of describing the diverse operations which shared the name 'acupuncture' (all of which he clearly considered to be secondary claimants to the name), Beck mentions 'a modification invented by Baunscheidt'. In this associated operation, forty shallow punctures were made into the body by an array of cutting needles set into a spring-loaded device with a flat circular head 'the size of a crown piece'. The bleeding wounds thus produced would be painted with an irritant 'which gave rise to an eruption like herpes'. It was the purifying suppuration of this eruption, rather than the actions of the needles themselves, which was meant to be therapeutic. Unpleasant as it may sound, this treatment had a wave of popularity; Beck noted, 'This was at one time in great repute as a quack remedy for all sorts of diseases.'[117] It was never accepted by orthodox practitioners, and I have not found any other contemporary British discussions of Baunscheidt's invention which liken it to acupuncture. A stronger connection between acupuncture and irregular medicine in this period could have been made in the area of male infertility where one shady group of practitioners promoted the idea that a needle inserted into the perineum or prostate could cure impotence, premature ejaculations, and 'nocturnal emissions'.[118] Again, I have found no direct link between pain-relief acupuncture and this particular therapeutic use of the needle. However, in the eyes of medical editors, even the purely coincidental link embodied in the shared instrument could have tainted the practice of pain-relief acupuncture with implications of quackery. Finally, Beck briskly cited the list of illnesses in which the mechanical effects of needling were considered to be beneficial; except for aneurisms, all involved the simple release of fluid. He clearly distinguished this group of operations from the use of the needle to relieve pain, in part by their use of different instruments: the former required the cutting-edged surgical needle, while the latter involved a round pointed needle with a cylindrical or grooved handle. This type of needle was designated as an 'acupuncture needle' and was sold under that name at a substantially higher cost than the surgical needles (possibly because their thinness required them to be more highly tempered).[119] Both of these orthodox, if perhaps uncommon, techniques were distinguished from the 'quackery' of Baunscheidtism.

Christopher Heath, also a member of the University College London faculty – in fact he was the Professor of Clinical Surgery and Beck's immediate superior – edited a *Dictionary of Practical Surgery* which was published in 1886. Acupuncture was given substantial coverage in this surgical compendium, with a three-page entry written by a surgeon named Chauncy Puzey. Puzey described the tools of acupuncture and its uses, initially distinguishing none of them as primary:

> Acupuncture may be performed according to circumstances, with a round-pointed or sewing, or with a cutting-edged or surgical needle. Its uses are various as a stimulant, as a counter-irritant; for the purpose of evacuating or dispersing fluids from various parts of the body; and the term may also be applied to the use of needles in the operation of tatooing, which is almost confined to ophthalmic practice.[120]

However, as the article continued, Puzey made it clear that 'acupuncture' was still conventionally used to describe the operation of needling for pain-relief: 'the term acupuncture is more generally applied to the use of needles for the relief of various painful affections, such as neuralgia and muscular pain, and stiffness consequent upon injury or chronic rheumatism'.[121] After describing the needles themselves and the proper insertion techniques for this form of acupuncture, he noted that it had been anciently used 'amongst the Orientals', and pointed out that in their hands, 'the effects are said to be marvellous' – unusually, he did not hint that 'marvellous' should be read as unbelievable. However, he did observe that 'the records of English surgery do not throw much light upon the subject ... its success has been by no means constant.'[122] Puzey referred to the recent operations made by Erichsen on sciatica (in the process noting that the patient's sensations determined the proper depth of the needle) and by two other surgeons on chronic muscular pain. Inevitably, he reiterated the persistent complaint that no mode of action had been determined and proven, but he also implicitly dismissed the linkage between Baunsheidtism and acupuncture proper: 'In the recent edition of Agnew's *Surgery* is shown an instrument for producing superficial acupuncture in the treatment of local pain; but here the effect may be clearly defined as counter-irritant.'[123] In addition, acupuncture had recently been tried in cases of fracture (where it was intended to stimulate circulation) and in the treatment of aneurism. It was this final mechanical use of the needle to reduce the pressure on swollen arterial vessels and prompt the growth of thick scar tissue at the delicate site which had captured Puzey's attention, and the remainder of the article was devoted to the topic.

Medical periodicals and the limits of local culture

> My object in communicating the following cases is to show the advantage of acupuncture as a remedy in the treatment of some forms of rheumatism and neuralgic affections.[124]
>
> John Tatam Banks, 1856

The close physical proximity within which Beck, Heath and Puzey worked is suggestive of the existence of self-perpetuating local cultures of acupuncture practice. University College Hospital in London may have been such a pocket, and the Leeds General Infirmary certainly was one. The articles that broke the journalistic silence on acupuncture confirm this picture. The first substantial and widely available periodical piece on acupuncture to appear after 1840 was written by John Tatam Banks and published in the *Lancet* in 1856. Banks reminded readers of his writings on acupuncture 'more than twenty years ago', and asserted that his opinion of acupuncture's powers had only been confirmed in the intervening decades: 'The experience of many years has confirmed my opinion of the value of acupuncture. I have ordered it in numerous cases – almost always with success – never (that I am aware of) with ill effects.'[125] Banks only mentioned China and Japan in passing, comparing their long history of successful acupuncture use with that of England where in his opinion, acupuncture was yet awaiting the 'unprejudiced trial it deserves'. Banks considered this English prejudice against the therapy to derive from two sources – scepticism of the unfamiliar and fear of the needle:

> The proposal to put a needle into the flesh as a remedial process is apt to excite a smile of incredulity, if not of ridicule; for few persons are inclined to think it probable that any benefit can arise from such a practice, and by many it is looked upon as a formidable remedy at best, while really it occasions but very trifling pain, and often scarcely any whatever.[126]

After this introduction, Banks turned to his argument in support of the needle, which he presented in two sections. First, he printed a letter from 'a surgeon and gentleman', describing how he fatigued and over-heated himself on medical visits, finally returning home in the chill of the night. This malign combination resulted first in fever, then incapacitating pain and stiffness of the chest and body – so much so that the surgeon, John Wrangham, thought he was bleeding internally, and called Banks to his assistance:

> [O]n your arrival, I rejoiced to find that you did not participate in the gloomy view I had taken of the case, but regarded it as one of a neuralgic and rheumatic character... You proposed *acupuncture*, which, I confess, I was inclined to oppose, having no predilection for *such* a remedy, and being somewhat an infidel as to any beneficial effects to be anticipated. You, however, were so confident (I may say so positive) that relief would follow the insertion of the needles, that I consented to the remedy...[127]

Naturally, acupuncture was effective, and Wrangham found himself once again able to breathe freely, but still experiencing muscular pain. ' "[E]ncouraged as I was by the great relief in my breathing" ', Wrangham is quoted as writing, ' "I now on my part anxiously wished for the introduction of other needles, along the course of the fibres of the several muscles affected." '[128] Banks inserted such needles, and his colleague was immediately relieved. The letter concluded in a tone of surprise and even wonderment – ' "I find it difficult to express in words the rapid and most extraordinary relief – as if by a charm – afforded by the insertion of the needles".'[129] Wrangham subsequently took up acupuncture himself.

Banks resumed the narrative; he offered his readers the by-now stock cautions – acupuncture could not cure rheumatism rooted in organic illness, nor treat cases where the site of pain was inflamed – before extolling its benefits in 'rheumatic or neuralgic pains, either acute or chronic . . . or those of an erratic kind (no matter where situated)'.[130] In particular, he asserted that acupuncture was not merely anagesic but curative. Then Banks turned to the second part of his argument, taking on one of the most esteemed medical resources, Dr Copland's *Dictionary of Practical Medicine*. Copland had scoffed at acupuncture as a therapy for rheumatism and neuralgia, and Banks quoted from his text before attacking it: ' "I have seen it much resorted to in several cases with success; but I am not aware of much permanent benefit having been produced by it. The practice has fallen into its deserved disuse." ' Banks's reply was sarcastic:

> Now this is a somewhat singular statement . . . Dr Copland admits that he has 'seen it resorted to in several cases *with success*' and yet says that 'the practice has fallen into its deserved disuse.' He says that he is 'not aware of much permanent benefit having been produced by it': it is surely but just to inquire whether he is aware, in the several cases in which he saw it resorted to with success, that permanent benefit was not produced by it? At all events, a remedy that, 'in several case has been resorted to with success' should be fairly tried; and found from repeated failures, to be of little utility before it receives condemnation from so high a quarter.[131]

From sarcasm, Banks turned to experience, reiterating his claim that acupuncture frequently produced permanent cures. He concluded, 'I should not have said thus much, did I not feel anxious to draw the candid attention of my professional brethren to a most valuable but much neglected remedy.'[132]

Banks apparently had not attached himself to any theoretical inter-pretation of acupuncture's effects, even after a additional twenty years of practice. However, in this stage of acupuncture's British sojourn, his reluctance to hypothesize was no longer shared by the majority of his colleagues. William Craig, a surgeon in Ayr, was treating a shoemaker troubled by a stubborn and extremely painful recurrent tumour. '[H]aving no confidence in any known method of treatment for the removal of pain', he was inspired to try acupuncture.[133] In 1859 Craig had written a text on the use of acupuncture for facial neuralgia in which he constructed a new version of a French theory of acupuncture's action. He argued '– That electric fluid and nervous fluid are identical'. Pain, at least in the case of *tic doloureux*, he believed to be caused by 'an accumulation of nervous fluid in the nerve affected'. The needles employed in acupuncture acted as conductors for this electro-nervous fluid, and thus 'the pain is instantly removed upon . . . insertion'.[134]

Craig drew an analogy between the stabbing and aching pains produced by his patient's tumour and those of acute neuralgia or tic doloureux, and decided to try the needles; his experiment was explicitly theory-driven. It also proved effective; Craig claimed that acupuncture's success 'was indicated at once by the countenance of the patient, which from being careworn and anxious became placid and cheerful'.[135] Clearly, the practitioner's necessary reliance on patient testimony to determine the success or failure of the treatment remained an undesirable feature of acupuncture, since Craig took pains to describe the physical signs of relief which he himself observed. However, the needle also had a strikingly visible effect on the malignant tumour; it was able 'to deprive it of its vitality'.[136] New tumours replaced the vitiated one, and the patient died, but he died free of pain. Craig was aware of the importance of multiple cases in the propagation of medical practices, but he argued that the urgent need for an effective analgesic made early publication imperative: 'The instant relief from agonies of the most excruciating character, and the deliverance from fears of their permanent continuance, is an object of the greatest importance to the patient, and cannot be uninteresting to him who seeks the best means of alleviating human suffering.'[137] He also commented on a Canadian report (published in the *Medical Times and Gazette* for 1865) of a successful acupuncture cure of tetanus, claiming that it was exactly the result predicted by his theory.

T. Pridgin Teale, of the Leeds General Infirmary was less inclined to theorize than Craig had been; his use of the technique was not prompted by any particular theoretical interpretation of the needle but was instead the fruit of local culture and networks. Acupuncture had been 'for years a favourite traditional practice at the Leeds Infirmary'. Pridgin Teale indeed seemed puzzled about acupuncture's previous obscurity:

> [I]t is my wish to record some facts concerning a method of treatment which, though boasting of great antiquity, and capable at times of doing good service, seems in a great measure to have dropped out of use, or at any rate to be at the present day but little employed or even known in many parts of the kingdom.[138]

Like all of the latter-day proponents of acupuncture, Teale was as anxious to limit the applications and expectation of acupuncture as he was to promote its use in the cases which he considered appropriate. His introduction to the case studies illustrated his dilemma: 'I do not profess that acupuncture succeeds in half, or even one third of the cases in which we use it; neither can I offer more than a conjectural explanation of its mode of action. When it does succeed, the relief it gives is often instantaneous.'[139] His caution, and that of his colleagues, echoes and reinforces Renton's early claim that acupuncture had been recklessly promoted in the past.

Teale was also cautious in advancing an opinion about acupuncture's mode of action. After presenting five case studies taken from the Infirmary's records, Teale turned almost reluctantly to the question of theory: 'When . . . we leave the facts and attempt to . . . explain how the remedy acts, it must be confessed that we tread on very uncertain ground. It is therefore with some diffidence that I venture . . . to suggest what appears to be a possible and reasonable explanation of results so definite and remarkable.'[140] Teale began by offering two categories of ailments for which acupuncture was suitable; the first, muscular disability, was uncontroversial and well-grounded in the British tradition of acupuncture. The second, which he himself called 'a somewhat artificial one', was more revealing of his attitude towards the therapeutic needle. He wrote, 'whenever a fixed pain has existed for some time and has resisted ordinary means of relief, general and local, I try acupuncture'.[141] In other words, acupuncture was decidedly not an 'ordinary means of relief', useful and traditional as it was. Rather it was again employed as a last resort, although one of which good things were expected at least half of the time. Using these categories and the evidence of his senses, Teale explored the question of *modus operandi*. He had observed that an aureole often formed around the needle in successful cases of acupuncture; he combined this observation with the Victorian interest in nutrition through a conviction that both muscular injury and chronic pain were the results of insufficient nutrition of the tissues. This starvation was caused by inadequate blood supply either to the tissue or to its nerves. From this core, Teale derived a theory by which to explain both the relief of pain and the restoration of muscular power and capacity. The needle, he argued, produced 'a temporary congestion and corresponding increase in the caliber of the vessels and in the blood supply', essentially priming the arterial pumps and thereby inducing an increased blood supply and proper bodily nutrition.[142]

Teale's article, like that of Tatam Banks, illustrates the power of formal and informal networks in preserving acupuncture through the years of its invisibility. Banks had learned of acupuncture in Paris during its British heyday; he continued to use it successfully for twenty years. Obviously, as an established user, Banks had no need of the periodical press to remind him of the needle's potential. Moreover, he was apparently converting neighbouring surgeons whenever he had opportunity to do so, largely by treating them (or their patients) successfully with acupuncture. The Leeds Infirmary too began to use acupuncture in the 1820s or 1830s, since Teale recorded his father's use of acupuncture on its wards; his father had preceded him in the post of Surgeon to the Infirmary, holding that office during the relevant period. Pridgin Teale Senior, like Churchill and the fortunate Thomas Martin, attended the United hospitals in London before returning to Leeds, and was probably exposed to acupuncture there. His colleagues at the Infirmary had also taken up acupuncture, and it had become established as orthodox, if not ordinary practice. The Infirmary in this period was progressive, and active in the pursuit of medical innovations – particularly those taking the form of new technologies. For example, in 1825, the Faculty agreed to purchase galvanic apparatus, and a year later decided to order a 'spirit vapour bath' as well. By 1832, the Infirmary had added to its list of desiderata a 'hydrostatic bed' which was purchased the next year.[143] Thus acupuncture use in the Leeds confirms both the importance of local networks, and the similar responses evoked by acupuncture and other instrument-centred medical innovations. Teale used acupuncture in his private practice as well, suggesting that the affluent as well as the destitute of Leeds were accustomed to the needle – certainly he recorded none of the scepticism so characteristic of more isolated acupuncture practices. Moreover, Teale noted that acupuncture had also been commonplace at the Birmingham Hospital. However, he reported that in Birmingham, acupuncture-use had died out as the reforming generation retired. Presumably, the apprenticeship and patronage system which had so benefited Teale's career also acted to prolong the use of acupuncture in Leeds by at least one more generation.

By the 1880s, articles on acupuncture were, if not a regular feature of the medical periodicals, at least not strikingly unusual. Acupuncture had returned to the public forum to the extent that, in 1885, two papers on the therapy were presented to the Annual Meeting of the British Medical Association in Cardiff. One was read in the section on Medicine, while the other occupied a parallel place in the Surgery division. The latter was given by J. Brindley James, a Member of the Royal College of

Surgeons, who presented a mechanical device designed to simplify acupuncture. After the meeting, he was persuaded by 'the solicitations of numerous professional friends' to publish his conference paper and a set of associated case studies as a short pamphlet entitled 'The Treatment of Lumbago and Rheumatic Pains by the Percusso-Punctator'.[144] His 'Percusso-Punctator' consisted of an array of five needles at one end of a spiral-grooved metal shaft, encased in an ivory handle; by twisting the shaft, the needles could protrude or be withdrawn to any controlled depth, and the shaft also could be electrified. James saw this tool as a means by which to promote the use of acupuncture in the treatment of lumbago, rheumatism, sciatica and other ailments:

> For years, I have obtained the most satisfactory results by acupuncture, effected through the means of a simple needle; but the very success of this system of treatment has induced me to seek and devise a means of facilitating, and thereby propagating, its application to a very wide extent.[145]

Brindley James, like his predecessors, was eager to communicate with other acupuncture users, and described himself as highly gratified, 'to hear my own views warmly advocated by no less an authority than Mr Macnamara' who presented several successful acupuncture cases to the South London district meeting of the Metropolitan Counties Branch of the British Medical Association.[146] Macnamara used 'simple acupuncture' to grant his patients 'complete immunity from pain', and James confirmed that he too had been satisfied with the results of acupuncture.

James then described forty-three cases in which he had used either his mechanical needles or a simple sewing-needle; these cases gave few details about the patients, only occasionally listing their ages. Their symptoms included persistent pain related to injury, nervous shock, rheumatism, hemicrania, brow-ague, dental neuralgia, sciatica, stiff neck and pleurodynia. In each of these complaints, he employed variously simple acupuncture, his percusso-punctator, and electro-puncture. But the majority of his studies described either of lumbago and vertigo.[147] He noted that he had not included rheumatic cases because he had successfully treated so many of them, and described one typical (and typically successful) example. In all, he recounted the cases of thirty-five men and eight women, all of whom left his surgery well, and on most of whom he had demonstrated the advantages of his gadget. He then turned to the unmechanized needle: 'previously to my adoption of my own invention ... the satisfactory results which had repeatedly followed the application of simple detached needles, and which had led me to seek in mechanical contrivance, a surer and easier method of resorting to the same, deserve a few words of notice'.[148] Enthusiastic as he was about his Percusso-Punctator, James clearly wished to promote acupuncture in any form, mechanical or manual. He told of three cases in which the simple needle 'acted like a charm', and concluded that he was 'deeply confident that its beneficial results will prove gratifying in the extreme'.[149] James discussed neither the history of acupuncture in Britain or Asia, nor any theory explaining its effects. He was essentially uninterested in the debate, and defined as his *modus operandi* the operation itself: 'It is of a most simple character: puncture, by means of needles, of the skin over the seat of the lesion.'[150]

In this respect in particular, James's article differed from that of his medical counterpart, Dr G. Lorimer of Buxton. Lorimer began with a long history of the needle, covering its ancient use in China and Japan; Ten Rhyne's transmission of it to the West; and the involvement first of the French Encyclopédists; then of their experimentalist successors, to whom he credited the first consideration of acupuncture 'to practical effect'. He cited Elliotson's *Cyclopedia of Practical Medicine* entry as the best source on the subject, combining as it did the history of acupuncture and the results of Elliotson's experience with the treatment at St Thomas's, but noted 'in modern times, it seems to be entirely disregarded and forgotten'.[151] Lorimer speculated on how this state of disregard had come about:

> Why acupuncture has passed into neglect it is not easy to explain. It may be that, in some conditions for which it has been used, it has been superseded by other and better means ... It may be that it has suffered at the hands of charlatans; and it may be, as had been suggested, that there is some general disinclination to use a remedial agent whose modus operandi cannot in some way be connected, analogically or otherwise, with that of the remedies which common use of universal experience has sanctioned.[152]

Lorimer's goal was to prove that this neglect was undeserved, and that acupuncture remained a 'prompt, efficient and reliable remedy'.[153] He first eliminated discussion of all the illnesses in which acupuncture, although effective, had been superseded by yet more effective means; sciatica cases, for example, had in the past been prime candidates for acupuncture therapy, 'but it [acupuncture] appears now to be eclipsed by the more formidable process of nerve-stretching'.[154] Instead, he chose to focus on rheumatism, for much the same reasons as those cited by James for its omission. His audience was presented with five successful and three failed cases of acupuncture use in rheumatic ailments. This selection was intended both to demonstrate the power of the needle and to verify, point by point, the conditions and limits which Lorimer had set upon the healing capacity of needling. He offered three cases of acupuncture for the relief of pain alone, a use for which he considered it less well adapted, then turned to a discussion of the *modus operandi*. Here he cited the various electrical and nervous explanations, but asserted that Teale's nutrition hypothesis was the most likely. Lorimer's interpretation of Teale's strongly reductionist explanation, however, reveals an underlying difference between the surgeon's and the physician's responses. Teale had considered that illness and disuse had reduced the size of the arteries supplying blood to the disabled region, rendering the 'mere act of will' impotent to force sufficient blood into the muscle and thus enable it to act. In his model, acupuncture, by creating a temporary congestion, expanded the vessels and allowed enough blood to build up that the muscle could be nourished and thus regain its activity. In other words, acupuncture produced a purely local and physical effect. Lorimer reported these views as 'not inconsistent with the conditions necessary for healthy muscular action'. But he added a caveat: 'How far, however, the *primum mobile* in the change is nervous agency, acting secondarily through the vascular supply of the affected muscles is open to question.'[155] Lorimer was not yet willing to give up either the systemic interpretation of these illnesses, or of acupuncture's effects.

In 1893, the first recognizably modern case study of acupuncture was published, fittingly enough in the *Lancet* which had long, if sporadically, supported the technique. It was authored by yet another Edinburgh-trained physician, E. Valentine Gibson, who had done his research as the senior medical resident at the Devonshire Hospital in Buxton, before moving to the more prestigious Victoria Infirmary in Glasgow. Although he made no mention of Lorimer, their geographical proximity is suggestive; Buxton was not a large town, though as a spa it supported more than the usual number of medical men. Gibson's article reported on 1000 cases of primary sciatica, but was focused on 'the treatment of one hundred cases by acupuncture'.[156] He presented his evidence without introduction – and with the assistance of (rudimentary) statistics. He observed, for instance, that he had eight male cases for each female, gave lists of age groups, and, more strikingly, created general categories into which every case was slotted, without reference to the idiosyncrasies or environment of the particular patient. Thus every case was defined solely by whether the patient was afflicted on the right, left or both sides of the body. Gibson reported his results with similar brevity:

> The results on discharge from the hospital of 100 consecutive cases of sciatica treated by acupuncture are as follows: 56 per cent were cured, 32 per cent were much improved, 10 per cent were improved, and in 2 per cent there was no improvement. These results I consider very satisfactory, considered the chronic nature and the severity of the majority of the cases... Acupuncture I consider very valuable.[157]

Like Wansbrough, Gibson critiqued an established medical source whose praise of acupuncture was lukewarm; Dr Gowers, an authority on the nervous system, had written in his textbook *Diseases of the Nervous System* that 'simple acupuncture along the course of the nerve... gives temporary relief, as does any superficial pain, but the cases are very few in which it has a permanent effect'. Gibson however, did not turn to sarcasm to voice his disagreement. Rather he challenged Gowers's acupuncture technique: 'I presume he must refer to cutaneous acupuncture, and not to acupuncture of the nerve itself, which was the method employed in these cases. The patient can always tell when the nerve has been pierced by pain shooting down the leg.'[158] Implicitly, Gowers was both old-fashioned and imprecise. Gibson's new technique, of course, bore little resemblance to either the British or the Asian traditions of acupuncture; however, in a passage further detailing his method, he made the intriguing note that 'even if the nerve is missed, the needle, passing in close proximity, must excert [sic] counter-irritation'.[159] This statement suggests that for all his apparent certainty, Gibson had seen effects which did not match his expectations, or fit his theoretical interpretation of acupuncture's mode of action.

As well as employing statistics, Gibson was innovative in considering the results of post-mortem studies of sciatic patients, and using their evidence (albeit acquired at second-hand) to support his interpretation of acupuncture: 'As I have had no opportunities of post-mortem examination I quote from Dr Gowers.' Gowers's research had found 'swelling and redness of the sheath and sometimes... small haemorrhages, and... slighter alterations in the interstitial tissue with secondary damage to the nerve fibers.' Gibson argued that if these findings revealed the true pathology of sciatica, 'the treatment by acupuncture is a rational one, more especially in the earlier stages of the disease, but even in the later stages puncturing the thickened nerve sheath may promote absorption'.[160] He gave no credit to the nutrition theory of acupuncture's action, turning instead to the pure reductionism of an earlier – and a later – era. Acupuncture, Gibson asserted, 'must give outlet, however small, to more or less of the exudation', and relieve the tension produced by the swelling of the nerve sheath and the blood vessels around it. Nonetheless, he had high hopes for the future use of the needle:

> [I]f every case of sciatica beginning acutely or subacutely were to be treated by absolute rest, together with acupuncture... and at the same time, any rheumatic or gouty tendency were treated by suitable remedies, I do not think there would be the number of chronic and relapsing cases that one so often sees.[161]

The end of the beginning: British acupuncture, 1890–1901

As the stigma which had attached to the 'quack' use of the needle in infertility and as a panacea faded out, both generations of British acupuncture users submitted articles on the technique to the periodical press. The first generation, those who had begun to use the needle in the 1820s, continued to promote it primarily with empirical evidence and pragmatic arguments. The second generation consistently linked the technique with theories explaining its effects – though not necesarily with the same ones. However, their theories were not drawn from the mainstream of British medicine, and certainly were not absorbed into it through the medium of articles on acupuncture. For all their orthodox credentials and even honours, the second ripple of acupuncture-users, practising mainly in the provinces, seemed a group slightly apart.

Despite their moderate success in returning the topic of acupuncture to the medical periodicals and public forums of the profession, acupuncture's late-nineteenth-century supporters were unable to restore the technique to its earlier prominence. In part, their failure was related to innovations in medical and surgical treatment of pain. Salycylic acid, uncomplicated, unthreatening and easily delivered in 'tabloid' form, had become the medical analgesic of choice for gout and rheumatism, while the technique of subcutaneous injections of anaesthetics – advocated by J. Brindley James alongside the Percusso-Punctator in the 1883 second edition of his pamphlet – was increasingly popular for sciatica.[162] Indeed, by 1908, J. Brindley James gave a paper on sciatica to the BMA Annual Conference in which he exclusively advocated injected sulphuric ether, re-writing his past practice at the same time. He noted: 'The sciatic nerve . . . has frequently been cut down upon or stretched or acupunctured for the relief of pain' with inconsistent results; he then described the injection treatment, asserting that 'I have always treated my cases of sciatica by this method.'[163] The only drawback to this therapy which he was willing to recognize was the irrational distaste for injections, 'which is felt because it is what the public call "an operation"', which would surely be removed by knowledge of the cures which were produced.[164] These treatments, if not yet completely explained by the theories and science of the day, were at least considered to be explicable; acupuncture was notoriously intransigent to theory. Furthermore, as medical science became more persuasive, therapies which resisted or seemed to resist experimental validation were threatened. Therapies which had long been considered 'irregular', like homoeopathy and hydropathy, were at least temporarily able to counter the authority of orthodox medical science with alternative systems of validation, developed in a parallel set of professional institutions, groups and publications. But acupuncture, as an orthodox but not mainstream therapy, depended on the activities of individual practitioners for publicity and validation. And, like J. Brindley James, individual practitioners moved on, turning to new therapies and more promising avenues to professional prominence. The needle lacked even the momentum produced by controversy – an odd minority practice was unlikely to receive the experimental attention given to the new diagnostic entities of bacteriology or the widespread but not yet stable practices of anaesthesia or antisepsis.

Between 1895 and 1970, popular and professional awareness of acupuncture as an available, therapeutic mode was minimal. After Gibson's 1893 article, the *Lancet* and the *British Medical Journal* maintained a complete silence on the subject of acupuncture for thirty years. Between 1930 and 1969, the *British Medical Journal* published six short and sceptical articles on acupuncture, while the *Lancet* continued its silence.[165] On the other hand, academic interest in Chinese medicine in general, including acupuncture, continued to increase in the first half of the twentieth century. Aided by a century of diligent collection and improved translation of Chinese medical manuscripts – efforts, ironically, spearheaded by the medical missionaries both to illustrate China's need for western medical aid, and to facilitate their own work in spreading the gospel of scientific medicine – physician-historians and orientalists constructed ever more sophisticated accounts of Chinese medical theory. Moreover, Chinese scholars, some trained in Britain, also began to produce histories of Chinese medicine and acupuncture. This work, much of it directed at popular as well as academic audiences, provided a more stable foundation for the modern transmission of information about acupuncture to Britain and the West.

Conclusions: Continuities in Cross-Cultural Medicine

> Memorandum on Certain Drugs, Formerly Used in Europe:...
> the enumeration as it stands will be sufficient to indicate the
> identity in several instances, and in others the similarity
> between the 'drugs' used 200 years ago in England and those
> at present employed in China. It is natural to assume that in the
> seventeenth as in the nineteenth century, the drugs enumer-
> ated were prescribed in accordance with particular theories in
> regard to the etiology and pathology of the diseases being
> treated. On this assumption the conclusion is justifiable that
> the theories of disease in this country and in the far East were
> very nearly, if not altogether identical.[1]
>
> Surgeon-General C. A. Gordon, Chinese Imperial
> Maritime Customs Service, 1884

Surgeon-General Gordon was an astute and sympathetic observer, both
of Chinese medicine and of his own time; under his editorship, the
annual medical reports of the Chinese Imperial Maritime Customs Service took on some of the functions of a learned journal for the amateur
botanists, naturalists and orientalists who served as medical officers for
the CIMCS.[2] The similarities which Gordon observed and reported
between Chinese medicine – its pharmaceuticals underpinned by ideas
of magical and systematic correspondence, and its nearly-humoral quality – and earlier western medical theory were valid ones.[3] Yet Gordon
couched his statement decidedly in the past tense; the beliefs shared by
earlier generations of Chinese and European physicians were now the
sole and undesirable property of the uninformed Chinese doctor. Gordon's words highlight the particular importance played by the timing in
the transmission of acupuncture. Humoral medicine, even underpinned

by scholastic anatomy as it was in seventeenth-century English and Scottish medicine, depended on vital principles very similar in their qualities and actions to the Chinese vital fluid, *qi*.[4] Had acupuncture been transmitted in conjunction with its theory in the first half of the seventeenth century, the practice might still have seemed alien, but the explanation behind it could easily have been assimilated. However, the technical expertise required to translate the specialized medical texts of China was rare even when Ten Rhyne was making his observations, and usually was confined to the Catholic missionaries. Acupuncture was transmitted to Europe with Ten Rhyne's approximations to Chinese medical theory, into a climate in which humoral and even hydraulic models of the body were being challenged by the proponents of mechanism and anatomy – in Britain, in particular, Baconians were gaining control over constructions of the body. The residue of theory which had adhered to the needle through the process of transmission was brushed aside by empiricism, and the separation between acupuncture's material and intellectual technology was complete.

At the beginning of the nineteenth century, an opportunity arose to re-unite the needle with its rationale. In Britain, galvanic, mesmeric and nervous models of the body all relied on a hypothetical active fluid – a fluid which again was strikingly similar to Chinese descriptions of *chi* – to unite the locally sympathetic tissues and organs into a functioning system. The resemblance between these models and the model of the body which explained acupuncture was remarkable. But by this time, acupuncture had already begun to take on a western form. Authors writing about the technique referred to an established canon of writing on acupuncture, all of which drew upon Ten Rhyne or Kaempfer as final authorities. The disjunction between practice and theory, material and cultural aspects of acupuncture had become fossilized.

'Acupuncture' and assimilation

Over the course of the nineteenth century, a therapeutic technique which its users called acupuncture was employed in Britain, from Ayr to Brighton, in private practice and public hospitals, and in rural as well as urban communities. The medical practice signified by the name 'acupuncture' was stable and consistent only in one respect: it always involved the insertion of a needle or needles into the body of the afflicted individual. In every other respect, 'acupuncture' was mutable. The needles could be intended, as in the traditional Chinese practice from which British acupuncture was so indirectly derived, to relieve through some unknown means the pain of a wide range of systemic complaints. However, the term 'acupuncture' could also signify a range of limited, mechanical processes involving the needle as a medical instrument. In its first British incarnation, as a means for relieving local pain, acupuncture was defined and promoted by a surgeon as a surgical technique; nonetheless, the needle came to be used by physicians, and appeared as frequently in medical dictionaries and compendia as in their surgical counterparts. Between 1822 and 1893, acupuncture was discussed in medical clubs and societies, and at professional meetings, and became established as an orthodox medical alternative in a range of illnesses.

Of the thirty acupuncture users about whom I was able to discover biographical information, half had some public consultancy work, although in most cases it was with dispensaries or local infirmaries, rather than the great urban hospitals. Fourteen were physicians (of whom at least eight trained in Scotland and two at Oxbridge). Among this group of acupuncturists were three Fellows of the Royal College of Surgeons in London and a similar number of Fellows at the Royal College of Physicians. Four acupuncture users are known to have travelled to Paris, and one explicitly noted his experiences there as the source of his interest in acupuncture. Within the orthodox profession, this was a diverse group; its members were bound together by a degree of moderate support for medical reform, and an equally temperate taste for acupuncture.

Although information about and use of acupuncture was widely dispersed, the practice of needling for the relief of pain was not necessarily common, even at the peak of its popularity in the late 1820s. Physicians and surgeons in Britain had unobtrusively employed acupuncture before it was reported in the British medical press – witness Mr Scott, whom Churchill cites as the first practitioner of acupuncture in England and the mysterious acupuncturists of Dumfriesshire, cited (but never named) by several contemporary authorities – and medical professionals continued to practise acupuncture after it fell from favour. Some, including surgeons at Leeds Infirmary, trained students or junior staff to use the technique and effectively perpetuated the use of acupuncture in their own local areas. Others convinced medical colleagues to employ the needle, or benefited from the stories of successfully cured patients. These personal networks and local traditions proved insufficient to permanently sustain the theory-free form of acupuncture practised in Britain, precisely because acupuncture users could not draw on the intellectual and emotional resources committed to theory. Instead, justifications of acupuncture depended almost entirely on its rate of empirical success, which in turn fluctuated wildly, depending on the practitioner and the disease being treated.

Acupuncture certainly did not fail to thrive in nineteenth-century British medical practice because of any violent opposition to its use. Nor, in this phase of its diffusion, had it brought with it any exotically indigestible understandings of the body. Early proponents constructed a niche for acupuncture in treating ailments which resisted established therapies, and relied on acupuncture's tactical value – initially to surgeons competing with physicians, and later to orthodox practitioners competing with medical irregulars – to promote its use. However, unattached to established theories, acupuncture made no claims on medical traditions or political loyalties, and supported none convincingly. In contrast to the eighteenth century, nineteenth-century discussions about acupuncture were not integrated with the lay or medical debates of the day, and if this protected its users from hostility (like that expressed by Wotton two centuries before), it also severed them from the more formal sources of credibility and professional support. This was the inevitable result of the strategy and rationales on which promoters of the technique relied to bring acupuncture into orthodox practice.

Acupuncture seems to have had little effect on the practice of medicine and surgery in nineteenth-century Britain. The use of acupuncture in Britain led to no general reconsideration of Chinese medical theory, nor to the adoption of the ideas underpinning its use – it would have been surprising if a technique which was associated at least as closely with France as with China in the minds of its users, and which had been stripped of its theoretical context a century earlier, had produced such effects. In fact, professional responses to Chinese medicine – and especially those of the medical missionaries – became more hostile as the century progressed despite the acceptance of acupuncture as a treatment modality in Britain.[5] This trend was to some extent balanced by the rise of scholarly curiosity and investigations into the subject, but neither development effected the decidedly low-key practice of acupuncture in the Victorian era. On the other hand, the therapeutic use of the needle in acupuncture does seem to have led to a general reassessment of the instrument's medical and surgical potential.

After Churchill's 1822 publication of the *Treatise on Acupuncturation* focused medical and surgical attention on the needle, it was rapidly taken up as an instrument to relieve oedema with reduced risk of infection or gangrene. Over the century, other uses for the needle were developed, some under the sheltering name of acupuncture. Lu and Needham, and Rosenberg, have proposed that acupuncture was important in the shift from the ivory lancet to the needle in vaccination; the technique of using a needle to press closed an artery or vein in surgery was first called acupuncture by its inventor; and an innovative acupuncture user claimed credit for introducing the subcutaneous injection of local anaesthetic.[6] Haller and others have noted that acupuncture users proposed the use of the needle also as an exploratory tool, and as a way to treat the acute dangers of aneurisms.[7] In the former process, the needle acted as a probe, to discover and sample tumours, cysts or other irregularities in the body's texture; in the latter, the needle was either manipulated to produce minor local bleeding and consequent clotting at the neck of the aneurismal sac, or to scratch the inner walls of the aneurism, causing them to produce more stable scar tissue.[8] Obviously, these operations have little in common with Chinese acupuncture, and share only slightly more with its western form as an empirical, surgical analgesic. Indeed, they express the very materialism which interpreted the complex Asian practice of acupuncture as simply and solely the therapeutic pricking of the body. The men who proposed these applications for the needle emphasized their novelty; however, they at least implicitly acknowledged a debt to the medical import in their choice of the name 'acupuncture'. To this limited extent, the nineteenth-century cross-cultural transmission of acupuncture can fairly be said to have affected the development of British medical and surgical practice. It is debatable, however, whether (at least after its British transformation), the status of acupuncture as a cross-cultural technique greatly altered the effects which the technique of analgesic needling would have had, were it an entirely native medical innovation.

Acupuncture after the National Health Service

Since the late 1960s and early 1970s, acupuncture has re-entered British medical culture and parlance, and (alongside other 'complementary medicines') been appropriated as an 'innovation' by both orthodox practitioners and medical consumers. In 1971, Chinese surgeons operated on the neck of a young male patient, who remained conscious throughout. When his wounds had been sutured, the patient sat up, had a glass of milk, spoke to his doctors and – much to their surprise – to a small group of observers, dressed himself and walked out of the theatre. His operation had been performed without chemical anaesthetic, and the only analgesia was provided by a set of steel needles, inserted into the patient's body. The observers in question were American medical men, invited by the China Medical Association to tour medical facilities in the newly open People's Republic of China. They reported their visit in the *Journal of the American Medical Association*, particularly emphasizing the remarkable use of needles – acupuncture – as analgesia even for major surgery. The Americans were swiftly followed by a group of British medical men, who published a very similar account of acupuncture in the *British Medical Journal* in early 1972, noting that 'in all cases the patients were conscious, fully co-operative, and appeared to suffer no pain...alternative methods of anesthesia were available but not used'.[9] Like their predecessors, the British medics were self-describedly 'astounded' by acupuncture anaesthesia. These medical delegations preceded Nixon's visit to China; however, theirs were not the first or necessarily the most influential descriptions of acupuncture to emerge from China in this period. The British and American broadsheets had beaten them to it by several months, with sensational and widely circulated accounts of a foreign correspondent treated under acupuncture anaesthesia while covering the Sino-American negotiations in Beijing. Both medical and popular reports, in their inception, their contents, and the reactions they strove to provoke, were intimately reflective of a rich political and cultural context.

The Vietnam War, the nuclear disarmament movement in the United Kingdom, the thalidomide disaster, and wide-ranging and widely read critiques like Rachel Carson's *Silent Spring* had provoked the disaffection of large segments of population in Britain from conventional authorities, and especially from medicine and science. This rejection of science as a purveyor of moral truths, combined with the environmental and feminist movements – and less progressively, with Britain's long-standing taste for orientalism – was reflected in rise of the New Age movement. For medicine, the New Age meant increasing consumer scepticism, a demand for professional accountability and responsiveness, and (at least ideologically) a call for simple or 'natural' medical treatments. The exotic drama – the spectacle – of acupuncture anaesthesia was certainly part of its immediate appeal, but so were the politics, and even the philosophy of care underlying the simple technology of surface-needling. This interplay of the exotic, the scientific and the pragmatic has shaped contemporary lay and professional responses to acupuncture, and thus mediated its diffusion.

There is, of course, a large and growing literature on the process and diffusion of medical innovation. While much of that work addresses individual techniques and specific technologies, medical innovation on a grander scale has not been ignored – for example, scholars have looked at the emergence of the hospital as a locus of care, at the use of computers in medicine, at changes in the roles of medical professionals and at changes in the funding and management of care.[10] Comparatively little attention, however, has been paid to medical innovations emerging from contexts other than those of the conventional laboratory, clinic or hospital. This is the more curious as the traditional model of medical diffusion – with knowledge moving from scientist to practitioner via formal public communication, and practitioners acting as individuals to assess, adopt, or reject that new knowledge or innovation – has been regularly challenged, adapted or dismissed altogether.[11] Barbara Stocking, in her study of innovation within the National Health Service, suggests that specific changes in patient care, whether technology-, knowledge- or structure-based are most likely to be initiated on the periphery, 'where the need is seen' – in other words, from 'service providers' rather than policy-makers. Of course, such initiatives must comply with policy mandates from the centre – calls for 'evidence-based medicine', or 'budgetary restraint', 'efficiency savings' or 'a focus on primary care'.

Stocking did not consider non-medical initiators in her study.[12] However, as social and cultural historians of medicine have demonstrated, patterns of medical innovation and knowledge-production do not comply with a sort of intellectual 'Central Dogma' of unidirectional flow, or necessarily evolve and function in the same way as innovations in manufacturing. We must therefore look beyond, as well as within, the conventional boundaries of biomedicine when seeking the catalysts of innovation. My research on the use of acupuncture in one NHS Trust hospital in Greater London suggests that for unconventional medicine, and particularly for cross-cultural practices, the process of innovation and diffusion is driven by forces on the periphery of professional medicine. This example of institutionalized acupuncture-use, while confirming the crucial role of service providers in prompting medical innovation, also demonstrates the central role of both medical consumers and heterodox practitioners in initiating change in the culture of medicine, even within the NHS.[13]

British consumers considering alternative medicine face substantial disincentives; until recently, the structure of the NHS made conventional treatment freely (if perhaps not readily) available, while consumers directly bore the cost of any therapeutic alternatives.[14] Nonetheless, British consumers turned to heterodox medicine in numbers which have steadily increased since the 1970s. Interest has grown among medical providers as well; in 1991, a survey of general practitioners revealed that 80 per cent were willing to refer patients to complementary practitioners, and by 1993, 65 per cent of District Health Authorities polled supported NHS-funding of at least some complementary therapies.[15]

Late in 1992, Stephen Dorrell, a minister in the Department of Health, announced that general practitioners who chose to refer their patients to complementary therapists would be allowed to pay for such treatment from their practice budgets. A few months later, in January 1993, a bill designed to establish osteopathy as a medical profession along the lines of dentistry received its second reading in Parliament. Also that month, perhaps in response to such proof of complementary medicine's growing legitimacy, staff at the Lewisham Hospital NHS Trust proposed an 'Acupuncture evaluation project' to be run by the hospital's own research wing. In a discussion paper, sponsoring staff argued that while good evidence existed for acupuncture's efficacy in pain relief, there was no such work explaining the variability in therapeutic effect from one patient or ailment to the next; and no data on the significance of traditional acupuncture points. The paper's author also criticized previous studies of acupuncture, in part for their failure to comply with the norms of biomedical research, but also – crucially – because such studies rarely assessed acupuncture as it was traditionally practised. In other words, this author acknowledged acupuncture as a complex of theory and practice, rather than just a technique or technology. The paper considered the difficulties inherent in applying the methodology of the random controlled trial to a therapy like acupuncture, and suggested an alternative. Instead of trying to eliminate or invalidate 'placebo' effects, the proposed trials would be designed to assess their extent and source. Three groups of patients would receive either conventional medicine alone, conventional medicine and traditional Chinese diagnosis, or conventional medicine, traditional Chinese diagnosis, and acupuncture. Such an experimental design not only eliminated the ethical dilemma of an untreated control group, but also distinguished between the effects of the medical encounter and the medical application. It also offered an implicit response to patient (and practitioner) dissatisfaction with the constrained structure of conventional medical encounters within the NHS, if only by testing whether the (admittedly idealized) traditional Chinese medical interaction – holistic, individualized and leisurely – was itself therapeutic.[16]

Nine months later, Lewisham Hospital's Health Services and Evaluation Unit formally proposed the establishment of a Complementary Therapy and Research Unit. The Unit was to provide complementary medicine for the hospital's internal market of patients and local client general practitioners, and to clinically evaluate the therapies on offer. After citing the 1993 reports of the British Medical Association and the National Association of Health Authorities and Trusts on the need for the NHS to 'embrace' complementary medicine, the proposal's introduction turns to the results of a poll of local GPs. Twenty-one per cent of the 71 respondents already employed complementary therapists within their practices (acupuncturists were the most likely to be based in a general practice), just over half referred patients to complementary therapists, and 65 said they would refer to the proposed unit.[17] The introduction then discussed how to integrate complementary therapies into the daily routines of the hospital, stressing institutional and structural issues. Tellingly, this semi-public document did not return to the point raised by the discussion paper about the importance of allowing complementary practitioners to follow traditional practices. Rather, it focused on restructuring complementary medicine along the lines of the orthodox profession, and essentially under its direction; referring physicians, for example, were to informed of any changes in their patient's status immediately, to receive full reports after first and last consultations, and to hold ultimate responsibility for the patient. Furthermore, all complementary therapists were required to participate in the research trials (with their heavy emphasis on standardization and documentation) on the basis that any 'unevaluated' forms of care paid for by the NHS should only be offered as part of research designed to assess them. A note that information about complementary medicine should be widely available to patients as well as practitioners appeared almost as an afterthought.[18]

Financially, the Complementary Therapy and Research Unit was to depend on two sources of funding. A research-driven strand would seek Medical Research Council funds for explorations of acupuncture, osteopathy and homoeopathy as 'evidence-based medicine'; while a demand-driven strand would draw upon purchasers among District Health Authorities and GP fundholders. Structural as well as ideological changes in the NHS were fundamental to emergence of such centres of innovative (and integrative) practice: the internal market, the fundholding GP, central demands for 'efficiency' and 'evidence-based medicine', and the ability of general hospitals to bid for research monies together gave Trust hospitals both incentive and means to fund such popular and potentially cost-cutting (or even fund-raising) measures. Only in the appendices were the therapies themselves, rather than their cost or popularity with local GPs, discussed. The authors defined each treatment, and offered a list of referral guidelines and counter-indications.

Perhaps unsurprisingly given this document's expected audience, acupuncture was described in terms of its Chinese origins and the mechanics of needling; the only marked difference from nineteenth-century definitions lay in a reference to its use as an anaesthesia.[19]

Lewisham's cautious proposal was successful, and in June 1994, Shadow Health Minister Dawn Primarolo opened the Complementary Therapy and Research Centre (CTC). The CTC shared a detached building with Lewisham Hospital's School of Nursing, and was in fact directed by Ann-Marie Brennan, a registered nurse and evaluated by another nurse, Janet Richardson, the manager of Lewisham's Health Service and Research Evaluation Unit. The involvement of senior nurses in establishing the CTC was far from coincidental; nurses have been active in promoting complementary approaches in the NHS. As Richardson and Brennan acknowledged in a 1995 article, 'In establishing the service, an opportunity was identified to create an innovative nursing role … [t]he Nurse Practitioner.'[20] In the CTC's first year, this role was biased towards administration; however, it was intended to develop into a clinical post drawing on the specialist complementary training of the Nurse Practitioner. The appeal of such a post needs little explanation, given the drive to professionalize and enhance the status of nursing within the NHS: 'The development of the Nurse Practitioner role will provide an opportunity to integrate nursing, therapy and research skills within a new domain of working, in which nurses are clearly taking the lead.'[21]

The CTC was funded by a block contract (for one year in the first instance) covering 200 new patient referrals per year, but allowing for extra-contractual referrals. In its first three months, it received 656 referrals.[22] The vast majority of those referrals were for back pain, arthritis and 'gynaecological problems', and most patients were referred to either acupuncture or osteopathy. Both of these are hands-on and time-intensive practices with lengthy diagnostic interviews. Both, as performances, differ strikingly from orthodox medicine, and (at least as practised within the NHS) both treat a limited number of chronic illnesses intransigent to conventional therapies. Although only a few patients presented as self-referred (indicative, perhaps, of the demographics of this working-class area), local GPs acknowledged the role of patient demand in their referral patterns. Structural changes within the NHS made the role of these patients visible – fundholding GPs were able to respond to their clients' requests, and their new financial clout enabled sympathetic parties within the local district hospital to make an appealing case for investigating complementary therapies.

In many ways, the creation of the CTC clearly resembles the innovations described by Stocking, Blume and others. The Complementary Therapy Centre was the end product of a process begun by the hospital's medical director (a consultant with interests in traditional African medicine) and a trained nurse working in its research evaluation unit. The medical director in question developed his own interest in acupuncture during the 1970s, precisely because of the media attention to the technique. He was therefore sympathetic to the proposal that the clinical utility and economic viability of acupuncture could be assessed at the hospital. Meanwhile, the nurse and researcher who instigated the project saw it as both intellectually and clinically valid, but also as a way of enhancing the status of nursing; her particular interest in acupuncture stemmed from centrality of the therapeutic interaction to the entire process. The documents surrounding the CTC repeatedly cite the new push for evidence-based medicine, and constantly return to the question of consumer demand for alternative therapies. This explicit acknowledgment of demand (and individual clients demanding as consumers, rather than as patients) – and the fact that it was acceptable to cite such demands as a justification for innovation at the clinical as well as the organizational level) set the Complementary Therapy Centre apart from other research projects conducted at the hospital. Thus where the January 1993 working paper indicated an interest in assessing the role of placebo effects and 'therapeutic interaction', and focused exclusively on acupuncture, the actual proposal submitted in autumn 1993 incorporated osteopathy and homoeopathy as well. This change was made explicitly because, with acupuncture, they were the most popular options among the hospital's client GPs. As part of its focus on therapeutic interactions, the original working paper stressed allowing complementary practitioners to treat within the parameters of traditional practice. This was probably the most radical innovation in the project, and in the final proposal, it disappeared.

After its first full year in operation, the CTC published an extensive evaluation of its research and services. This report made a number of recommendations about developing an integrated service with other departments of the hospital. Here complementary medicine is expanding much like any orthodox innovations.

> The nature of the referrals suggests that a number of patients may also be receiving treatment from other departments within [the hospital] . . . the service should begin to consider how an integrated service, based on packages of care, can be provided for the full benefit of

patients. For example, patients attending the pain clinic might benefit from acupuncture, or patients referred to the orthopedic or phy-
siotherapy department may benefit from osteopathy in addition to
the 'conventional' therapy...[23]

The report argued strongly for a fuller integration of complementary
medicine into the NHS and greater collaboration with other colleagues.
It also noted that the research model set up for complementary medicines could usefully be extended to orthodox units where random controlled trials have not been carried out to evaluate procedures in
accordance with evidence-based medicine directives. In other words,
an innovation related to controlling unconventional medicine was to
be imposed upon orthodox departments and therapies. As the author of
the report noted:

> The move to consumerism signals a new phenomena in medicine,
> and could have a major impact on the provision of non-conventional
> therapies in the NHS. Technological and therapeutic advances have,
> to a great extent, determined the nature of procedures and treat-
> ments available to patients. Thus the provision of health care has
> been 'profession driven'. There is however, a growing recognition
> that patients are in fact 'partners' in healthcare, that their participa-
> tion in the process of treatment can influence compliance and out-
> come, and that ultimately, 'their views are the ones that matter.'[24]

The CTC was ambitious and perhaps unpolitic in suggesting that acupuncture and osteopathy be integrated into the repertoire of Lewisham
Hospital's orthodox clinics, and that the same standards should be
applied to the assessment of conventional and complementary therapies. Such changes clearly complied with central directives to eco-
nomize and to employ only 'evidence-based medicine'. That these
policies have not always been wholeheartedly welcomed by the medical
profession may partially explain the local Health Authority's decision to
close the CTC in 1997, despite continued over-subscription to its services, and much positive media attention.

Continuities in the cross-cultural transmission of medical knowledge

This contemporary phase of transmission and response largely repro-
duces eighteenth- and nineteenth-century attempts to integrate the

needle into medical practice. The three waves of transmission examined here share certain characteristics. First, each phase of transmission followed a surge of popular interest in China – the first two waves coincided with Chinoiserie fads – sparked by increased access to Chinese culture. Second, each transmission of acupuncture occurred in the context of medical contention and change. Third, in each case, the process of transmission came about and was inflected by the actions and interests of comparatively few members of the medical profession, acting more or less individually.

The most striking similarity, of course, is that as in each preceding surge in the visibility of British acupuncture, the current popularity of the technique comes at a time of contention, competition and change in the structures and values of orthodox western medicine. When Ten Rhyne described acupuncture in the seventeenth century, he was addressing a community in metamorphosis; mechanism and Baconian empiricism were challenging Galenic/scholastic approaches to disease, and consumers were increasingly choosing between these systems as they selected a medical practitioner.[25] Unsurprisingly, the stresses of transition first became evident where the established Galenic therapeutic modes had long been found wanting; gout was one such point. The proper treatment of gout was also a subject which traditionally had divided physicians from surgeons. However, neither conventional surgery nor orthodox medicine offered a safe cure for the disease, and the rigid resources of mechanism had not provided a compelling explanation for gout. The needle certainly offered the former, and as interpreted by Ten Rhyne, acupuncture provided evidence for the latter; these conjoined attributes enabled the alien technique to impinge upon the European debate.[26] Gout, where three tributaries of change in European medicine converged, remained a visible locus for acupuncture use until salicylic acid became a consensus therapy in the second half of the nineteenth century.

In Regency Britain, as well, competition between surgeons and physicians, and contention between modernizing reformers and status quo traditionalists created a niche in which acupuncture could flourish.[27] From the mid-century on, surgeons and physicians were competing less strongly with each other than with irregular practitioners; as an orthodox alternative, acupuncture retained some tactical value in the struggle. In particular, orthodox practitioners employed the needle for exactly the chronic painful and nervous conditions – for example sciatica, arthritis, neuralgia, hypochondria and migraines – which attracted so much irregular attention. In the absence of more successful or more flexible analgesics, acupuncture could be publicly defended on empirical grounds even in a climate which increasingly valued theoretical and experimental justifications. With the *fin-de-siècle* resolution of the conflict between regular and irregular systems, brought about largely by the overwhelming success of both new models of disease and new modes of clinical knowledge, acupuncture lost its strategic value. Simultaneously, the marginalization of the diseases for which acupuncture had been considered appropriate eroded its visibility. In contemporary medicine, these trends have been reversed. The demographic shift towards the aged has created a vast market for geriatric medicine, while quality-of-life concerns have returned chronic and painful conditions to medicine's centre-stage. The ways of conceptualizing and treating illness which have dominated orthodox western medicine throughout the twentieth century are themselves proving unsatisfactory in relation to these re-discovered diseases. Orthodox and alternative therapies are once more in competition, and practitioners from a range of medical systems are contending for authority and credibility, as well as business.

Paradoxically, the fact that traditional Chinese medicine (along with westernized versions of Chinese practices) is among the contenders for acceptance and status in contemporary British medicine indicates both the similarity of this transmission process to those which preceded it, and its essential difference. As in the eighteenth and nineteenth centuries, acupuncture has risen to prominence in conjunction with a spike of interest in China and Chinese culture. Each time, this interest has coincided with increased access to China (or at least information about China) – in the twentieth century, this expansion took the form of the re-opening of China to the West in the late 1960s and early 1970s. However, where previous generations of the British public knew China primarily through its exported material goods, modern British consumers from across the economic and educational spectrum can also purchase – indeed are deluged with – the products of Chinese culture.[28] Western medical professionals continue to separate the technique and technology of acupuncture from the theories underlying it, but the integrated original form is actively practised as well, both within the institutions of orthodox British medicine and beyond them. Nineteenth-century scholars of China laid the foundation for this increased access through their collection and translation of medical and philosophical writings. Although Lewisham's conservative proposal to its Health Authority showed little insight into traditional acupuncture practice, both the original discussion paper and their first

'Operational Policy' booklet revealed a greater awareness of acupuncture's complexity:

> Acupuncture is a branch of Traditional Chinese Medicine which uses an integrated system of diagnosis incorporating tongue observation and pulse reading. Developed over two thousand years ago, it is based on the principle that the stimulation of certain areas of the body influences the functioning of the physical organs.[29]

Neither homoeopathy nor osteopathy was described in terms of its history or cultural origins, and for none of the other therapies was the diagnostic process detailed – at least to that extent, orientalism clearly still clings to the medical imagination. Although this brief definition does not address the theories underlying Chinese medicine, it accurately reflects traditional Chinese understandings of acupuncture's mode of action. Without the work of medical orientalists, twentieth-century interpreters like Felix Mann in Britain and G. Soulié de Morant in France could not have re-presented a domesticated and yet essentially intact acupuncture to their respective medical establishments.[30]

The successful transmission of apparently discrete medical practices seems surprisingly dependent upon the inclusion of intellectual technology (expertise) along with its material counterpart. The importance of medicine's immaterial components can be assessed by comparing the results of the deliberate missionary effort to export western medical practices to China with those of the haphazard and fragmented transmission of acupuncture in the opposite direction. European medical missionaries rapidly learned that transplanting medical practices from one cultural milieu to another required the conjoined forces of empirical success and compelling explanations for those results. Early efforts had demonstrated that merely providing free medical care, and relying on word of mouth to spread the news of its empirically proven superiority could lead to undesirable results. In response to high levels of eye-infection and blindness among the Chinese, the first British medical facility in China was dedicated to treating ophthalmic disease – literally as well as figuratively, it was an endeavour designed to bring light to the people.[31] However, although the facility soon acquired a widespread reputation, it was not necessarily the one intended by its founders. One popular Chinese woodcut showed a man in Chinese dress, having his eyes gouged out by two European medical men. Western medicine was subsequently exported as a package: as they demonstrated medical and surgical procedures, the medical missionaries also expounded their theories, their models of illness and health, and their views of the body. Even the forms of western medical education were exported. Doctors ordered anatomical charts and chemistry sets from Britain, and reported their classes to be eagerly awaiting the arrival of these tools. A major effort followed to translate basic western medical and scientific texts into Chinese. Western medicine flourished in China, although it neither replaced traditional Chinese medicine, nor remained untouched by its new environment.

In contrast, acupuncture was transmitted to and propagated in Europe piecemeal through the actions and interests of individual western medical practitioners. After the first wave of transmission, these practitioners had no knowledge or experience of acupuncture as it was practised in its native context, nor even, as non-Chinese speakers, access to the basic texts. Since neither the Chinese nor the Japanese made any attempt to export their medicine to Europe, and, at least initially, the few medical men who spoke Chinese were fully occupied in translating western medicine into Chinese, the theory remained obscure and distant from the practice of acupuncture, and acupuncture failed to take root in British practice.

In contemporary Britain, the Chinese conception of the body as a dynamically balanced whole, and the style of medical practice which developed in response to that conception have greatly contributed to acupuncture's popularity. The radically different theoretical constructions of traditional Chinese medicine have become an attractive feature for medical consumers disenfranchised by orthodox bio-medicine. Similarly, some medical practitioners have adopted needling in part because the slower pace and more intimate tone of acupuncture-practice better matches their own sense of how the doctor–patient interaction should be. In this stage of acupuncture's British history, the efforts of individual practitioners remain central to the spread of acupuncture within the medical profession. However, the return of the conditions which have favoured medical boundary-hopping in the past – contention, competition, change and the availability of alternatives (both to consumers of medicine and to providers of medical care) – suggests that medical consumers will regain at least a measure of their former control over the therapeutic encounter. Desperate and dissatisfied consumers, and in particular consumers for whom Chinese theory is at least as accessible and comprehensible as the abstruse doctrines of bio-medicine, have little to lose from adopting exotic alternatives. Assimilation, even for a technique which remains unexplained by western theories, no longer necessarily entails dilution.

In the eighteenth and nineteenth centuries, acupuncture, as a candidate for the East-to-West transmission of medical expertise and understanding, was neither a natural choice, nor an inevitable failure. Its complex and often obscured relationship with Chinese models of the body (themselves first inaccessible and then unassimilable in the West) and its apparently irreducible strangeness to western eyes rendered acupuncture immediately suspect. However, its contemporary success in Britain demonstrates that these inhibiting factors were products of particular historical moments and events – as indeed may be its current popularity. In studying the cross-cultural transmission of medical knowledge, it is occasionally necessary to peel away the visible generalities – the clashing models of health or the body, the alien quality of individual therapies, or of moments in time – to discover the specificities which drive each episode of assimilation or rejection; from those points a more reliable, if less predictable pattern may emerge.

Notes

Introduction

1 Joseph Banks, quoted in P. J. Marshall, 'Britain and China in the Late Eighteenth Century', in Robert A. Bickers (ed.), *Ritual and Diplomacy: the Macartney Mission to China, 1792–1794* (London, 1993), 11–29, at p. 25.

2 See, for example, David Arnold, (ed.), *Imperial Medicine and Indigenous Societies* (Manchester, 1988); David Arnold, *Colonizing the Body: State Medicine and Epidemic Disease in 19th Century India* (Berkeley, 1993); John Z. Bowers, *Western Medical Pioneers in Feudal Japan* (Baltimore, 1970); E. Richard Brown, 'Exporting Medical Education: Professionalization, Modernization, and Imperialism', *Social Science & Medicine*, 13A (1979): 585–95; Roy MacLeod and Milton Lewis (eds), *Disease, Medicine and Empire: Perspectives on Western Medicine and the Experience of European Expansion* (London, 1988); Bridie Andrews and Andrew Cunningham, *Western Medicine as Contested Knowledge* (Manchester, 1997).

3 Of course, the influence of Arabic science and medicine on the early development of the European disciplines clearly models East-to-West cross-cultural transmission of knowledge; see A. A. Khairallah, *Outline of Arabic Contributions to Medicine* (Beirut, 1946); Lucien Leclerc, *Histoire de la Médecine Arabe: Exposé Complet des Traductions du Grec; les Science en Orient, leur Transmission à l'Occident par les Traductions Latines* (New York, 1961); Y. A. Shahine, *The Arab Contribution to Medicine* (London, 1976); and Donald Campbell's classic (if naturally dated), *Arabian Medicine and its Influence on the Middle Ages* (London, 1926). For information on the Asia trade and European consumption, see S. A. M. Adshead, *Material Culture in Europe and China, 1400–1800: the Rise of Consumerism* (Basingstoke, 1997); Chaudhuri, *The Trading World of Asia and the English East India Company, 1660–1760* (Cambridge, 1978); Neil McKendrick, John Brewer and J. H. Plumb, *The Birth of a Consumer Society: the Commercialization of 18th Century England* (London, 1982); H. B. Morse, *The Chronicles of the East India Company, Trading to China, 1635–1834* (Oxford, 1926 9); John Willis Jr, 'European Consumption and Asian Production in the Seventeenth and Eighteenth Centuries', in John Brewer and Roy Porter (eds), *Consumption and the World of Goods* (London, 1993), 143–77. And of course there are honourable exceptions among historians of economic botany in particular: see, for example, Lucille Brockway, *Science and Colonial Expansion: the Role of the British Royal Botanical Gardens* (London, 1979); Clifford Foust, *Rhubarb: the Wondrous Drug* (Princeton, 1992); Nick Jardine, James Secord and Emma Spary (eds), *Cultures of Natural History* (Cambridge, 1996); David Miller and Peter Reill (eds), *Visions of Empire: Voyages, Botany and Representations of Nature* (Cambridge, 1996).

4 I frequently use the term 'Asian medicine' in this volume; this is not because Japanese, Chinese, Ayurvedic and other regional medical systems are monolithic or undifferentiated. Rather, it is because the national variants of acupuncture and of explanations for it were essentially undifferentiated by European

commentators. Early reports tended to use 'Japanese' and 'Chinese' more or less interchangeably in their discussions of medical ideas, remedies and techniques. In nineteenth-century Britain, China and Japan were perceived very differently, and some supporters of acupuncture attempted to exploit this difference by associating the needle with the well-regarded Japanese instead of the 'vain and ignorant' Chinese. However, in most cases, no such distinction was made and in none did the distinction perceptibly affect responses to acupuncture. In cases where the difference is important, I have used national adjectives advisedly; elsewhere, I have generally used the term 'Asian' or (and particularly in relation to medical theory) reverted to 'Chinese'.

5 Louis LeComte, *A Compleat History of the Empire of China: Being the Observations of Above Ten Years Travels through that Country. . . A New Translation from the Best Paris Edition* (London, 1739), 229.

6 John Barrow, *Travels In China. . . In Which It Is Attempted To Appreciate The Rank That This Extraordinary Empire May Be Considered To Hold In The Scale Of Civilized Nations* (London, 1804), 298.

7 Barrow, *Travels*, 304–5.

8 Barrow, *Travels*, 304–5.

9 Sir George Staunton, *An Authentic account of an Embassy from the King of Great Britain to the Emperor of China. . . Taken Chiefly from the Papers of His Excellency the Earl of Macartney. . . and of Other Gentlemen in the Several Departments of the Embassy* (London, 1797), Vol. 3, 380.

10 See Robert Fortune, *A Journey to the Tea Countries of China* (London, 1852) and Lucile Brockway's short summary of these events in *Science and Colonial Expansion*, 27–8. See also Lisbet Koerner, 'Purposes of Linnaean Travel: a Preliminary Research Report', in David Miller and Hans Reill (eds), *Visions of Empire: Voyages, Botany, and Representations of Nature* (Cambridge, 1996), 117–52.

11 David Mackay, 'Agents of Empire: the Banksian Collectors and Evaluation of New Lands', in Miller and Reill (eds), *Visions of Empire*, 38–57. See also the Reeves MSS, BL Add. MSS, 33982 and 35262; and Wellcome MSS, 5217, items 34–6.

12 Breton, *China: Its Costume, Arts, Manufacture, &c. Edited Principally from the Originals in the Cabinet of M. Bertin; with Observations Explanatory, Historical, and Literary, by M. Breton*, 2nd edition, Vol. 1 (London, 1812), 8.

13 *The Universal History* (London, 1759), 647. See also William Wotton, *Reflections upon Ancient and Modern Learning* (London, 1694); Louis LeComte, *A Compleat History*, 229

14 Breton, *China*, Vol. 3, 15.

15 Edward Said, *Orientalism: Western Concepts of the Orient* (London, 1978); and *Culture and Imperialism* (London, 1993). Subsequent studies and critiques include Homi Bhabha, *The Location of Culture* (London, 1994); Lisa Lowe, *Critical Terrains: British and French Orientalisms* (Ithaca, 1991); Ranajit Guha, *Subaltern Studies: Writings on South Asian History and Society* (Delhi, 1981) and subsequent edited volumes; Bryan Turner, *Orientalism, Postmodernism and Globalism* (London, 1994) offers a useful tripartite definition of 'orientalism': 'First, orientalism can be regarded as a mode of thought based on a particular epistemology and ontology which establishes a profound division between the Orient and the Occident. Second, orientalism may be regarded as an

academic title to describe a set of institutions, disciplines and activities usually confined to Western universities which have been concerned with the study of oriental societies and cultures. Finally, it may be considered as a corporate institution primarily concerned with the Orient' (96). 'Medical orientalism' partakes primarily of the first two aspects. For examinations of medicine, imperialism and orientalism, see David Arnold, *Colonizing the Body*; Elisabeth Hsu, 'The Reception of Western Medicine in China: Examples from Yunnan', in Patrick Petitjean, Catherine Jami and Anne Marie Moulin (eds), *Science and Empires: Historical Studies about Scientific Development and European Expansion* (London, 1992), 89–101; A. Kumar, *Medicine and the Raj: British Medical Policy in India, 1835–1911* (London, 1998); Kabita Ray, *History of Public Health in Colonial Bengal, 1921–1947* (Calcutta: 1998). For a glimpse of the impact of the colonies on British culture, see the essays in Julie Codell and Dianne Sachko Macleod, *Orientalism Transposed: the Impact of the Colonies on British Culture* (Aldershot, 1998); Paul Greenhalgh, *Ephemeral Vistas: the Expositions Universelles, Great Exhibitions and World's Fairs 1851–1939* (Manchester, 1988); John MacKenzie (ed.), *Imperialism and Popular Culture* (Manchester, 1986); Harriet Ritvo, *The Mermaid and the Platypus and Other Figments of the Classifying Imagination* (Cambridge, MA, 1997).

16 G. Tradescant Lay, *The Chinese as They Are: Their Moral, Social and Literary Characters; A New Analysis of the Language; with Succinct Views of Their Principal Arts and Sciences* (London, 1841); W. H. Medhurst, *The Foreigner in Far Cathay* (London, 1872). These are only two of the more popular examples.

17 John Davis, *The Chinese: a General Description of the Empire of China and Its Inhabitants* (London, 1837), 272–6.

18 Davis, *The Chinese*, 277, emphasis in the original.

19 Samuel Kidd, *China, or, Illustrations of the Symbols, Philosophy, Antiquities, Customs, Superstitions, Laws, Government, Education, and Literature of the Chinese* (London, 1841), 364.

20 Frederic Henry Balfour, *Leaves from my Chinese Scrapbook* (London, 1887), 64–5.

21 Balfour, *Chinese Scrapbook*, 65.

22 Remarkably little has been written on the transmission of acupuncture to Europe in the late seventeenth and early eighteenth centuries, or on the western reception and use of acupuncture in the two centuries that followed this first transmission. See Lu Gwei-Djen and Joseph Needham, *Celestial Lancets: a History and Rationale of Acupuncture and Moxa* (Cambridge, 1980) for a valuable survey of early European literary and philosophical responses to Chinese medicine. After the re-opening of China in the 1960s and 1970s, a flurry of short articles on acupuncture appeared in medical and popular journals and in western newspapers. See for example, John Z. Bowers and Robert Carrubba, 'The Doctoral Thesis of Engelbert Kaempfer: "On Tropical Diseases, Oriental Medicine and Exotic Natural Phenomenon"', *Journal of the History of Medicine and Allied Sciences*, 25 (1970): 270; Carrubba and Bowers, 'The Western World's First Detailed Treatise on Acupuncture: Willem Ten Rhijne's *De Acupunctura*', *Journal of the History of Medicine and Allied Sciences*, 29 (1974): 371–98; John Haller, 'Acupuncture in Nineteenth Century Western Medicine', *New York State Journal of Medicine*, 73 (1973): 1213–21; Elisabeth Hsu, 'Outline of the History of Acupuncture in Europe', *Journal of Chinese Medicine*, 29 (1989): 28–32; George Rosen, 'Lorenz

Heister on Acupuncture: an Eighteenth Century View', *Journal of the History of Medicine and the Allied Sciences*, 30 (1975): 386–8; Dorothy Rosenberg, 'Wilhelm Ten Rhyne's De Acupunctura: an 1826 Translation', *Journal of the History of Medicine and the Allied Sciences*, 34 (1979): 81–4; Dorothy Rosenberg, 'Acupuncture and U.S. Medicine: a Socio-Historical Study of the Response to the Availability of Knowledge' (PhD dissertation, University of Pittsburgh, 1977); Mike Saks, *Professions and the Public Interest: Medical Power, Altruism and Alternative Medicine* (London, 1995). For a limited but useful examination of the French response to acupuncture, see Daniel Geoffroy, *L'Acupuncture en France au XIXe siècle* (Paris, 1986). The history of acupuncture in China has, of course, been studied far more extensively. As a sample of the many approaches, see Wong Chimin and Wu Lien-te, *History of Chinese Medicine: Being a Chronicle of Medical Happenings in China from Ancient Times to the Present Period* (Tientsin, 1932); William Morse, *Clio Medica XI: Chinese Medicine* (New York, 1934); Paul Unschuld, *Medicine in China: a History of Ideas* (Berkeley, 1985); and of course, Lu and Needham, *Celestial Lancets*. For more on change and the rhetoric of stasis in modern Chinese medicine, see R. C. Crozier, *Traditional Medicine in Modern China: Nationalism and the Tensions of Cultural Change* (Cambridge, MA, 1968) and Bridie Andrews, 'The Making of Modern Chinese Medicine, 1895–1937' (PhD dissertation, University of Cambridge, 1996). I am not a sinologist, so I have shamelessly pillaged the work of those more diligent scholars to identify Chinese individuals, texts and places, wherever such information might be useful to myself or readers. The origins of this borrowed erudition are identified as opportunity arises. In many cases, that origin will be Lu Gwei-Dgen and Joseph Needham's *Celestial Lancets*. To those readers who desire a more detailed treatment of Chinese culture and history in this period of contact and exchange, I recommend the work of Jonathan Spence.

23 For example, John Elliotson left abundant correspondence, now in the National Library of Scotland, documenting his better known predilections for mesmerism, phrenology and trousers, but I found no reference to his interest in and use of acupuncture. Admiral Sir James Coffin, who seems to have browbeaten his medical attendants into treating him with acupuncture, left behind only papers relating his term in Parliament and a minor naval invention. Such silences, though frustrating, do reveal the degree to which acupuncture was absorbed into orthodox medical culture.

24 The situation in France was slightly different, as at least one nineteenth-century doctor and acupuncture-user (Sarlandière, 1825) did accept certain aspects of Chinese theory, although he subsequently modified the practice of acupuncture. His interpretation had no effect on British acupuncture and very little on French uses of the needle, but his modification of needling technique – electrifying the needles in situ – was adopted in some cases.

1 Expectations and Expertise

1 G. Tradescant Lay, 'Minutes of the First Annual Meeting of the Medical Missionary Society in China', in *The First and Second Reports of the Medical*

Missionary Society in China: With Minutes of Proceedings, Hospital Reports, &c. (Macao, 1841), 9.

2 P. J. Marshall, 'Britain and China in the Late Eighteenth Century', in Robert A. Bickers (ed.), *Ritual and Diplomacy: the Macartney Mission to China, 1792– 1794* (London, 1993), 11–29. For more on European sinology, see Ming Wilson and John Cayley (eds), *Europe Studies China* (London, 1995) and T. H. Bartlett, *Singular Listlessness: a Short History of Chinese Books and British Scholars* (London, 1989).

3 See Michael Adas, *Machines as the Measure of Men: Science, Technology, and Ideologies of Western Dominance* (Ithaca, 1989), 21–69; Jonathan Spence, *The Search for Modern China* (New York, 1990), 132–6. As David Arnold suggests in *Colonizing the Body: State Medicine and Epidemic Disease in 19th-Century India* (Berkeley, 1993), India presents a variation upon this theme of enthusiasm, followed by exploration (physical, cultural and intellectual), followed by devaluation. The subcontinent's fall from grace occurred in two stages, interrupted by the brief Orientalist rediscovery of Indian science and mathematics in the late eighteenth and early nineteenth centuries (43–58).

4 Assessing the accuracy of the claims themselves is properly the province of historians of Chinese medicine and will not be addressed here; readers curious about this point may wish to consult the compendious work of Paul Unschuld, Nathan Sivin, Joseph Needham and colleagues, and other sinologists.

5 For a compelling analysis of the mission as an exercise in natural history, see James Hevia, *Cherishing Men from Afar: Qing Guest Ritual and the Macartney Embassy of 1793* (London, 1995), 84–115.

6 John Barrow, *Travels In China, Containing Descriptions, Observations, And Comparisons, Made And Collected In The Course Of A Short Residence At The Imperial Palace Of Yuen-Min-Yuen, And On A Subsequent Journey Through The Country From Pekin To Canton. In Which It Is Attempted To Appreciate The Rank That This Extraordinary Empire May Be Considered To Hold In The Scale Of Civilized Nations* (London, 1804), 3.

7 Barrow, *Travels*, 3. Barrow (1764–1848) was Comptroller to the Embassy, a post he gained through the patronage of Sir George Staunton. He subsequently became Second Secretary to the Navy, and was crucial in shaping its policies of exploration in Africa and the Arctic, as well as government responses to China.

8 J. L. Cranmer-Byng, *An Embassy to China: Lord Macartney's Journal, 1793–1794* (London, 1962), 3–58. For another lucid account of this period with a somewhat different perspective, see Jonathan Spence, *Search*, Chapters 6 and 7; D. E. Mungello, *Curious Land: Jesuit Accommodation and the Origins of Sinology* (Honolulu, 1989).

9 Sir George Staunton, *An Authentic Account of an Embassy from the King of Great Britain to the Emperor of China . . . Taken Chiefly from the Papers of His Excellency the Earl of Macartney . . . and of other Gentlemen in the Several Departments of the Embassy*, Vol. 1 (London, 1797), 41.

10 Staunton, *Account*, Vol. 2, 60–1.

11 Traditional Chinese diagnosis did not (and does not) rely solely on the pulse; like their European counterparts, Chinese physicians were meant to employ all five senses – including listening to their patients – in diagnosis. The pulse

did have particular importance in prognosis. Two recent and intriguing essays exploring this subject are Shigehisa Kuriyama, 'Visual Knowledge in Classical Chinese Medicine', in Don Bates (ed.), *Knowledge and the Scholarly Medical Traditions* (Cambridge, 1995), 205–34; and Francesca Bray, 'A Deathly Disorder: Understanding Women's Health in Late Imperial China', in Don Bates (ed.), *Knowledge and the Scholarly Medical Traditions* (Cambridge, 1995), 235–50. Both note that Chinese medicine in this period is known predominantly through the scholarly writings, and hence is itself the medicine of the elite.

12   Staunton, *Account*, Vol. 2, 61–2.

13   Barrow, *Travels*, Vol. 3, 345–6.

14   For a detailed description of the British clinical encounter of the day, see Dorothy Porter and Roy Porter, *Patient's Progress: Doctors and Doctoring in Eighteenth-Century England* (Stanford, CA, 1989), especially Part II.

15   Staunton, *Account*, Vol. 2, 62, my emphasis.

16   Barrow, *Travels*, Vol. 3, 346.

17   For more on the creation of cultured bodies, see Ludmilla Jordanova, *Sexual Visions: Images of Gender in Science and Medicine Between the 18th and 20th Centuries* (Madison, 1989); Catherine Gallagher and Thomas Laqueur (eds), *The Making of the Modern Body: Sexuality and Society in the Nineteenth Century* (Berkeley, 1987); Thomas Laqueur, *Making Sex: Body and Gender from the Greeks to Freud* (Cambridge, MA, 1990).

18   Anne Digby documents a medical community dominated by the (affluent) patient. See Anne Digby, *Making a Medical Living: Doctors and Patients in the English Market for Medicine, 1720–1911* (Cambridge, 1994). Also consider William Bynum, 'Health, Disease and Medical Care', in Roy Porter and George Rousseau (eds), *The Ferment of Knowledge* (Cambridge, 1980), 212–13; and Porter and Porter, *Patient's Progress*, Part I. Bray, 'Deathly Disorder', 237–9, notes that the medical attendants of the elite in China, as in Europe, were necessarily (and for the same reasons – lower social status and financial dependence) listeners and negotiators, whatever their occasional claims to the contrary.

19   See Porter and Porter, *Patient's Progress*, Part II, 5; Bynum, 'Health, Disease and Medical Care', 213; and for more detailed accounts, including discussion of the changes in this patient-centred practice, Stanley J. Reiser, *Medicine and the Reign of Technology* (Cambridge, 1978), Chapter 1.

20   It was (and is) actually somewhat unusual for a traditional Chinese practitioner to rely exclusively on the pulses, and especially not to also examine the tongue. However, it was certainly the ideal of Chinese physic to be free from the patient self-reporting, to derive diagnostic information entirely from the body, more or less unmediated. In a medical culture where mind and body were resolutely considered as one entity, the subjective experience of the patient would theoretically be legible from the body's state.

21   The western understanding of the pulse exemplified by Barrow and Staunton was constructed over the course of the eighteenth century; at the beginning of the century, responses to Chinese pulse diagnosis were considerably different. In Chinese anatomy, *qi* flows through channels linking related organs; the multiple pulse of Chinese medicine is taken from different points

on the arm where such channels pass close to the surface of the body. Thus the Chinese pulse carries complex information about an interrelated system.

22 Barrow, *Travels*, Vol. 3, 345–6.

23 J. B. DuHalde, *A Description Of The Empire Of China And Chinese-Tartary, Together With The Kingdoms Of Korea And Tibet: Containing The Geography And History (Natural As Well As Civil) Of Those Countries. Enrich'd With General And Particular Maps, And Adorned With A Great Number Of Cuts. From The French Of P. J.B. Du Halde, Jesuit: With Notes Geographical, Historical, And Critical; And Other Improvements, Particularly In The Maps, By The Translator* (London, 1738–41), 124.

24 DuHalde, *A Description*, 184.

25 DuHalde, *A Description*, 183.

26 R. James, *A Medicinal Dictionary; including Physic, Surgery, Anatomy, Chymistry, and Botany, in all their Branches Relative to Medicine . . . and an Introductory Preface, Tracing the Progress of Physic, and Explaining the Theories which have Principally Prevail'd in All the Ages of the World* (London, 1743), viii, my emphasis.

27 See Porter and Porter, *Patient's Progress*, 160–1; Charles Rosenberg, 'The Therapeutic Revolution: Medicine, Meaning, and Social Change in 19th Century America', in Judith Leavitt and Ronald L. Numbers (eds), *Sickness and Health in America: Readings in the History of Medicine and Public Health*, 2nd edition (Madison, 1985), 39–52; Nicholas Jewson, 'The Disappearance of the Sick-Man from Medical Cosmology, 1770–1870', *Sociology*, 10 (1976): 225–44.

28 James, *Medicinal Dictionary*, viii.

29 James, *Medicinal Dictionary*, viii.

30 See Michael Barfoot, 'Brunonianism Under the Bed: an Alternative to University Medicine in Edinburgh in the 1780s', in William Bynum and Roy Porter (eds), *Brunonianism in Britain and Europe* (London, 1988), 22–45. As well as being medically radical, Brunonianism was associated with social and political radicalism. It is worth noting that Gillan's expressed ideas on fever, his enthusiasm for bleeding, and in particular his criticism (echoed by Staunton) of a Chinese healer who prescribed stimulants to a British sufferer, are anti-Brunonian.

31 Chris Lawrence, 'Cullen, Brown, and the Poverty of Essentialism', in Bynum and Porter, *Brunonianism*, 1–21. Indeed, Lawrence points out that *both* Brown and Cullen portrayed themselves as 'supporter[s] of the view that progress in medicine was to be achieved by employing fundamental philosophical principles in order to arrive at a general explanation graced by causal simplicity' (7). Thus, the absence of system would have condemned Chinese medicine in either case. For further discussions of Edinburgh, the Scottish Enlightenment, and the conflicting roles of Cullen and Brunonianism, see Bynum and Porter, *Brunonianism*; L. Stephen Jacyna, *Philosophic Whigs: Medicine, Science and Citizenship in Edinburgh, 1789–1848* (London, 1994); Chris Lawrence, 'Ornate Physicians and Learned Artisans: Edinburgh Medical Men, 1726–1776', in Bynum and Porter (eds), *William Hunter and the Eighteenth-Century Medical World* (Cambridge, 1985), 153–76.

32   Staunton, *Account*, Vol. 3, 55.
33   Hugh Gillan, 'Mr. Gillan's Observations on the State of Medicine, Surgery, and Chemistry in China', in Cranmer-Byng, *An Embassy to China*, 279–90, at p. 281.
34   Staunton, *Travels*, Vol. 3, 55–7.
35   See Gillan, 'Observations', 281.
36   Gillan, 'Observations', 281, my emphasis.
37   Gillan, 'Observations', 282. Given the extraordinary precision with which the appropriate points for feeling the various pulses were located in Chinese medical theory, one wonders if this did more harm than good – the sight of an alleged doctor fumbling around on the arms of a very influential patient may have destroyed any chance that remained of western medicine being taken seriously at the Chinese court after the frequent illnesses of the Ambassadorial party.
38   Gillan, 'Observations', 280–1.
39   See Sir John Floyer, *The Physician's Pulse-Watch; or, an Essay to Explain the Old Art of FEELING the PULSE, and to improve it by the Help of a Pulse-Watch . . . To Which is Added, An Extract out of Andrew Cleyer, Concerning the Chinese Art of Feeling the Pulse* (London, 1701).
40   Gillan, 'Observations', 280. Cranmer-Byng adds to this the gloss that DuHalde translated the Mo ching, written circa 10th century AD, and published this translation in 1735, with his *Description*.
41   Gillan, 'Observations', 282.
42   Staunton, *Account*, Vol. 3, 57–8.
43   The Chinese physicians were of course diagnosing a disease of disordered *qi*. For the purposes of this article, *qi* can be defined as 'universal vital force'; in traditional Chinese medicine, *qi* was understood to be composed of two elements, yin and yang, which circulated throughout the body, the former with the blood and the latter in a second, more ethereal, system of channels. However, it is clear from their respective texts that both Staunton and Gillan chose to interpret *qi* as referring to a gas in the modern sense rather than in the older sense of *spiritus*. This interpretation seems to have been based primarily on the treatment proposed by the Chinese: acupuncture – which Gillan and Staunton both presented as an attempt to create a physical outlet for a physical substance – 'opening passages for its escape, directly though the parts affected'.
44   Staunton, *Account*, Vol. 3, 378–9.
45   Rosenberg, 'Therapeutic Revolution', 39.
46   Floyer, *Physician's Pulse-Watch*, 2.
47   Leslie Stephen and Sidney Lee (eds), *The Dictionary of National Biography, from Earliest Times to 1900*, Vol. 7 (Oxford, 1949–50), 346–9.
48   Floyer, *Physician's Pulse-Watch*, 2.
49   Floyer, *Physician's Pulse-Watch*, title page.
50   Floyer, *Physician's Pulse-Watch*, 229.
51   Floyer, *Physician's Pulse-Watch*, 230–1.
52   Floyer, *Physician's Pulse-Watch*, 232.
53   Floyer, *Physician's Pulse-Watch*, 'Preface', 2.
54   Floyer, *Physician's Pulse-Watch*, 'Preface', 3.
55   Floyer, *Physician's Pulse-Watch*, 336.

56 Floyer, *Physician's Pulse-Watch*, 355.
57 Floyer, *Physician's Pulse-Watch*, 368.
58 Floyer, *Physician's Pulse-Watch*, 247.
59 For more on the shared roots of Chinese and western medicine, see Shigehisa Kuriyama, 'Interpreting the History of Bloodletting', *The Journal of the History of Medicine and Allied Sciences*, 50 (1995): 11–46.
60 Gillan, 'Observations', 283–4.
61 Gillan, 'Observations', 284.
62 Gillan, 'Observations', 279.
63 Gillan, 'Observations', 283.
64 For details on this contest and its origins, see Porter and Porter, *Patient's Progress*; M. Jeanne Peterson, *The Medical Profession in Mid-Victorian London* (Berkeley, 1978); L. Stephen Jacyna, *Philosophic Whigs*, especially his chapter on pathology.
65 Gillan, 'Observations', 279.
66 See Paul Unschuld, *Medicine in China: a History of Ideas* (Berkeley, 1985), as to the extent and visibility of this culture; for comparisons, see also the reports of the French Jesuits in Fr. Jean-Baptiste DuHalde (ed.), *Lettres Édifiantes et Curieuses Écrits Des Missions Etrangères* (Paris, 1702–76). An English translation had appeared by 1742, and a second edition of this translation came out in 1761.
67 Gillan, 'Observations', 284
68 Staunton, *Account*, Vol. 3, 92–3.
69 Gillan, 'Observations', 283. There is no good way to verify that Ho-Shen indeed spoke these words; however, if Gillan was determined to show his procedures in the most positive light, he certainly could have attributed to Ho-Shen a less ambiguous turn of phrase.
70 Staunton, *Account*, Vol. 3, 60–1, my emphasis.
71 Incidentally, he was roundly criticized for his presumption in publishing an account; it was considered inappropriate for one of his low status to have views on China, and particularly to criticize his employer's handling of the political and diplomatic issues involved.
72 Aeneas Anderson, *A Narrative of the British Embassy to China, in the Years 1792, 1793, and 1794; With Accounts of the Customs and Manners of the Chinese; and a Description of the Country, Towns, Cities, &c.*, 2nd edition (Dublin, 1796), 275.
73 Staunton, *Account*, Vol. 3, 92–3.
74 Staunton, *Account*, Vol. 3, 274–5.
75 Chinese metallurgists actually did produce a silvery alloy at this time which was slower to tarnish than silver. It is possible that Eades may have heard of this metal, called *pe-tung* by the Chinese and white copper or *patkong* by the Europeans, through the reports of Catholic missionaries; alternatively, he may have seen specimens of the metal which had come to England in trade. For a somewhat more detailed contemporary description of this alloy, see Gillan, 'Observations', 292–3 and ff. 149.
76 John Harley Warner, 'The Idea of Southern Medical Distinctiveness: Medical Knowledge and Practice in the Old South', in Leavitt and Numbers, *Sickness and Health*, 53–70. This essay focuses on the arguments for southern distinctiveness; however, Warner amply describes some of the medical beliefs which underpinned that argument and the deliberate and explicit creation of

distinct medical cultures in rural and urban, northern and southern, European and American contexts.

77 Todd Savitt, 'Black Health on the Plantation: Masters, Slaves, and Physicians', in Leavitt and Numbers, *Sickness and Health*, 313–30.

78 M. Jeanne Peterson, *The Medical Profession*; Judith Walkowitz, *Prostitution and Victorian Society: Women, Class, and the State* (Cambridge, 1980). See also Laura Engelstein, 'Morality and the Wooden Spoon: Russian Doctors View Syphilis, Social Class, and Sexual Behavior, 1895–1905', in Gallagher and Laqueur, *Making Sex*, 169–208, where the effects of context on the interpretation of disease entities is discussed.

79 Charles Rosenberg, 'Social Class and Medical Care in 19th Century America: the Rise and Fall of the Dispensary', in Leavitt and Numbers, *Sickness and Health*, 273–86 for a case study of the role of economics in creating distinct medical modalities.

80 This encounter also sheds some light on Jewson's hypothesis that consensual diagnosis and treatment formed the foundations of Enlightenment medicine. Anderson's acceptance of the Chinese mode of diagnosis, in which he read the patient's body, rather than calling upon the patient's own account, and subsequently prescribed for him without entering into any dialogue, lends some weight to Porter's suggestion that consensus-medicine was the province of the relatively elite consumer, and that doctor–patient relationships took other forms when the patient bore other status. See Jewson, 'Disappearance' and Jewson, 'Medical Knowledge and the Patronage System in 18th Century England', *Sociology*, 8 (1974): 369–85; Roy Porter, 'Laymen, Doctors, and Medical Knowledge in the Eighteenth-Century: the Evidence of the *Gentleman's Magazine*', in Roy Porter (ed.), *Patients and Practitioners: Lay Perceptions in Pre-Industrial Society* (Cambridge, 1985), 283–314.

81 Dr Dinwiddie, quoted in William Jardine Proudfoot, *Biographical Memoir of James Dinwiddie...Embracing Some Account of his Travels in China and Residence in India* (Liverpool, 1868), 87.

82 See Spence, *Search*; Jonathan Spence, *Memory Palace of Matteo Ricci* (New York, 1984); Unschuld, *Medicine in China*; Mungello, *Curious Land*.

83 Adas, *Machines*. See particularly his Introduction and Chapter 1, and 231–4, 248–52, on the relationship with China.

84 See Proudfoot, *Biographical Memoir*, 47 and his rebuttal of Barrow's tale of 'the Emperor's Favourite Draughtsman'.

85 Barrow, *Travels*, 306–7. It seems very likely that Barrow borrowed this insight from DuHalde, *Description*, Vol. 2, 124.

86 Barrow, *Travels*, 354. It is, of course, suggestive that the passage specifies an 'Edinburgh Surgeon'. Clearly, the distinction between the Scottish medical mode and its more traditional equivalents was evident by the close of the eighteenth century.

87 Staunton, *Account*, 379–80.

88 Staunton, *Account*, 380.

89 Proudfoot, *Biographical Memoir*, 53.

90 Proudfoot, *Biographical Memoir*, 53, italics in the original.

91 Proudfoot, *Biographical Memoir*, 46. The Chinese reaction to these gifts, and the implications which that reaction held for subsequent events in

Anglo-Chinese relations, has been discussed at length by Spence, *Search*; and mentioned in Bickers, *Ritual and Diplomacy*; and Hevia, *Curious Land*.
92 Proudfoot, *Biographical Memoir*, 47–8. In the end, despite his fascination with China's grand engineering projects, and small technical innovations – for Dinwiddie prepared notes on everything from the canal network to their remarkable skill at cutting glass – the Chinese reaction to western science led Dinwiddie to despair of China: 'The extreme jealousy, added to the extreme ignorance of the Chinese, will prevent our visiting the manufactures, & c. Nothing but conquest by some polished nation will ever render this a great people. The prejudices are invincible. Ask them whether the contrivers and makers of such curious and elegant machinery must not be men of understanding, and superior persons. They answer – "These are curious things, but what are their use? Do the Europeans understand the art of Government as equally polished?"'
93 Cranmer-Byng and Trevor Levere, 'A Case Study in Cultural Collision: Scientific Apparatus in the Macartney Embassy to China, 1793', *Annals of Science*, 38 (1981): 503–25.
94 Anderson, *A Narrative*, 177.
95 Cranmer-Byng, *An Embassy*, 96.
96 Anderson, *A Narrative*, 133.

2 The Needle Transfixed

1 The orthographic abandon with which eighteenth-century authors treated proper nouns meant that at least two and often multiple accepted spellings existed for the names of foreign places and individuals. Thus, the man I have chosen to call Wilhelm Ten Rhyne, following the title page of his *Dissertation de Arthritide*, was also called Ten Rhijne and ten Rhyne, while the *De Arthritide* is catalogued in the British Library under the surname Rhyne. Similarly, Kaempfer has also been Kempfer, Kemper, Kampfer and Kæmpfer. Chinese and Japanese names for both places and people were even more liberally varied, and I have generally left them in the form which the authors of each text used – noting their shared referent in places where wildly deviating spellings might obscure it. Western transliterations of Chinese medical terms have in some cases rendered the original terms untraceable, even through context-clues.
2 Sir John Floyer, *The Physician's Pulse-Watch; or, an Essay to Explain the Old Art of FEELING the PULSE, and to improve it by the Help of a Pulse-Watch... To Which is Added, An Extract out of Andrew Cleyer, Concerning the Chinese Art of Feeling the Pulse* (London, 1701), 337.
3 For more information on the consumption of Indian, Chinese and Japanese products in Europe at the time, see John Willis Jr, 'European Consumption and Asian Production in the Seventeenth and Eighteenth Centuries', in John Brewer and Roy Porter (eds), *Consumption and the World of Goods* (London, 1993), 133–47; K. N. Chaudhuri, *The Trading World of Asia and the English East India Company, 1660–1760* (Cambridge, 1978); S. A. M. Adshead, *Material Culture in Europe and China, 1400–1800* (New York, 1997). For general and lively accounts of consumption and the development of markets in

eighteenth-century Britain, see Neil McKendrick et al., *The Birth of a Consumer Society: the Commercialization of 18th Century England* (London, 1982).

4   Sir George Staunton, *An Authentic Account of an Embassy from the King of Great Britain to the Emperor of China . . . Taken Chiefly from the Papers of His Excellency the Earl of Macartney . . . and of other Gentlemen in the Several Departments of the Embassy*, Vol. 3 (London, 1797), 57–8.

5   See Introduction.

6   J. S. Forsyth, *The New London Medical and Surgical Dictionary* (London, 1826).

7   John Z. Bowers, *Western Medical Pioneers in Feudal Japan* (Baltimore, 1970); Lu Gwei-Djen and Joseph Needham, *Celestial Lancets: a History and Rationale of Acupuncture and Moxa* (Cambridge, 1980). Ten Rhyne (1647–1700) was in Japan 1673/4–5/6.

8   'An Account of a Book, viz. Wilhelmi ten Ryne M. D. & c. . . . I. dissertatio de Arthritide. 2. Mantissa Schematica: 3. de Acupunctura. . . . Londini in 8vo 1683'. *Philosophical Transactions*, 13 (1683): 222–35, at p. 222.

9   The Japanese term for the policy of closing Japan to the outside world.

10   Wilhelm Ten Rhyne, *Dissertatio de Arthritide: Mantissa Schematica: De Acupunctura: Et Orationes Tres: I De Chymiae et Botaniae Antiquitate et Dignitate, II. De Physionomia, II De Monstris* (London, 1683). The work was published under the auspices of the Royal Society.

11   'An Account', *Philosophical Transactions*, 222–35.

12   Ten Rhyne himself cites Martinus Martinius, a Jesuit in China, whose work was entitled *Sinicae Historiae decas prima, res a gentis origine ad Christum natum . . . gestas complexa* (Munich, 1658); and Jacob Bontius's enormous *Historia Naturalis & medicas Indiae orientalis*, Book 5 (Amsterdam, 1658). The latter he quoted on the use of acupuncture for colic, but cautioned that Bontius was incorrect on several points. Andreas Cleyer – who went to Japan twice in the 1680s and with whom Ten Rhyne had at least one medical dispute – published several remarkable diagrams of the acu-tracts and their associated viscera in 1682, but his *Specimen medicinae Sinicae* focused on Chinese theories of the pulse and their diagnostic techniques, rather than on acupuncture and moxibustion. Lu and Needham, *Celestial Lancets*, 276–85 add to this list of predecessors, but acknowledge Ten Rhyne's position as the first major European spokesman for acupuncture.

13   Roy Porter and G. S. Rousseau, *Gout: the Patrician Malady* (New Haven, 1998), 48–67.

14   Richard Ingram, *The Gout. Extraordinary Cases in the Head, Stomach and Extremities; with Remarks and Observations on the Various Stages of the Disorder – the Rheumatism – the Disease Commonly Called the Scurvy – the Nature and Formation of External and Internal Chalk-Stones – and Considerations Proving Gout the Immediate Parent of Jaundice, Dropsy and Stone. . . . to which is Prefixed an Essay Pointing out the Progressive Symptoms and Effects and the Reasons Why the Gout was not heretofore Regularly Treated and Cured* (London, 1768), 9.

15   Porter and Rousseau, *Gout*, 71–124.

16   John Douglas, *A Short Dissertation on the Gout. Wherein the Universal Fear of Doing Anything to Ease or Cure It (Instilled in People's Heads by Both Ancient and Modern Writers) will be Proved to be a Mere Bug-Bear, a Groundless Supposition, a Vulgar Error, & c. and a Safe Method of Relieving the Most Violent Pains, Shortening the Fit, and Lengthening the Intervals, will be Proposed, and Confirmed by Several*

Cases (London, 1741). Douglas was a surgeon, and a Fellow of the Royal Society.

17 Among the many roughly contemporary descriptions and diatribes on both gout and quack medicines, the following are notable for their clarity (and frankness) in uniting the two issues: T. Garlick, *An Essay on the Gout... The Remedies, both Internal and External Faithfully Publish'd in English, without Reserve, for the Benefit of all such as now do (or hereafter may) Suffer by that Disease...* (London, 1729); Dale Ingram, *An Essay on the Cause and Seat of the Gout: in which the Opinions of Several Authors are Considered, and Some External Operations Recommended* (Reading, 1743); and Richard Ingram, *The Gout.*

18 Lu and Needham, *Celestial Lancets* offer some speculations on the chronology of this transmission. See also William Temple, of whom more later. Note that the reasons Temple gives for trying moxa would not support acupuncture in the same way, and might even agitate against its adoption.

19 Robert Carrubba and John Z. Bowers, 'The Western World's First Detailed Treatise on Acupuncture: Willem Ten Rhijne's De Acupunctura', *Journal of the History of Medicine and Allied Sciences*, 29 (1974): 371–98, at p. 371.

20 'An Account', *Philosophical Transactions*, 222–3.

21 Carrubba and Bowers, 'De Acupunctura', 377–8. Ten Rhyne refers to Iwanga Sokaa and Motogi Shodayu, respectively. Ten Rhyne described the questions as 'Bothersome trifles, to be sure.' Carrubba and Bowers note that Ten Rhyne's answers to some 150 questions like 'Why do you feel only the left pulse?' and 'How do you differentiate the Yang-type carbuncle and the Yin-type carbuncle?' were later published in *Zen-seishi-Tsuiwa*, Vol. 1, Book 2 (1680), 372.

22 Carrubba and Bowers, 'De Acupunctura', 377–8.

23 Although the writings of Descartes in the 1670s induced the birth of mechanism, medical writers and theorists were not immediately converted. See, for further discussion, Eric Carlson and Meribeth Simpson, 'Models of the Nervous System in Eighteenth Century Psychiatry', *Bulletin for the History of Medicine*, 43 (1969): 100–15.

24 The system of acupuncture is based on the idea of a circulation of this substance, *qi* (or *chhi*, *chi*, and in Japan, *ki*). Unfortunately, that term has a rich concretion of meaning in Chinese, representing a constellation of qualities, properties and entities physiological and otherwise. It is virtually impossible to translate, a difficulty the results of which would become evident in European interpretations. As Lu and Needham put it, 'we do not yet know how best to translate chhi... We even doubt whether there could ever be a justified one-word European translation... we said that chhi was something like pneuma, i.e. subtle spirits, tenuous matter, something resembling air, or a gas or vapour, but also something which could have the character of radiant energy like radioactive emanation, or x-rays, or very highly penetrating particles'. In modern medical texts, as Lu and Needham also observe, the terms 'energy' and 'vital energy' have become standard translations. Lu and Needham, *Celestial Lancets*, 16, Note a.

25 A less crucial set of mistranslations derived from the fact that *qi* is a balance of *yin* and *yang* components/energies, the former of which circulates with the blood and the latter of which passes only through these vessels.

26 Carruba and Bowers, 'De Acupunctura', 382–3. Ten Rhyne did, however, admit that classical physicians had been similarly loose in their terminology,

citing Rufus of Ephesus, another recently rediscovered Classical medical writer.

27   Carruba and Bowers, 'De Acupunctura', 383, my emphasis. The reference to 'cutting' is puzzling, as dissection was virtually unknown in Japan at this time.

28   'An Account', *Philosophical Transactions*, 230–1. Since this passage, in various translations and versions, appeared in almost every subsequent account of Chinese 'anatomy', I have quoted this contemporary English translation of Ten Rhyne (as opposed to Carrubba and Bowers's smoother modern version) at some length. I have, however, removed most of the measurements of length, leaving only samples to give a flavour of the paradox which must have struck British readers. It seems likely that Kee Miak is a transliteration of *chi mo*, and Rack Miak of *lo mo*.

29   Carrubba and Bowers, 'De Acupunctura', 376.

30   Carrubba and Bowers, 'De Acupunctura', 376.

31   Shigehisa Kuriyama, 'Visual Knowledge in Classical Chinese Medicine', in Don Bates (ed.), *Knowledge and the Scholarly Medical Traditions* (Cambridge, 1995), 205–34 offers an intriguing discussion of the place and use of the medical gaze in Chinese anatomy, as well as some comparisons with western anatomy and medical vision.

32   Lu and Needham suggest that the engraver was responsible for these alterations; I have found no good evidence on which to base a claim either way.

33   Carrubba and Bowers, 'De Acupunctura', 376.

34   Carrubba and Bowers, 'De Acupunctura', 376.

35   Carrubba and Bowers, 'De Acupunctura', 374–5.

36   See Londa Schiebinger, *The Mind Has No Sex? Women in the Origins of Modern Science* (Cambridge, MA, 1989); C. Gallagher and T. Laqueur (eds), *The Making of the Modern Body: Sexuality and Society in the Nineteenth Century* (Berkeley, 1987); Thomas Laqueur, *Making Sex: Body and Gender from the Greeks to Freud* (Cambridge, MA, 1990).

37   Carrubba and Bowers, 'De Acupunctura', 375.

38   Carrubba and Bowers, 'De Acupunctura', 375.

39   Carrubba and Bowers, 'De Acupunctura', 396.

40   Carrubba and Bowers, 'De Acupunctura', 375.

41   Carrubba and Bowers, 'De Acupunctura', 376.

42   Carrubba and Bowers, 'De Acupunctura', 377.

43   Carrubba and Bowers, 'De Acupunctura', 377.

44   Carrubba and Bowers, 'De Acupunctura', 391–3.

45   Carrubba and Bowers, 'De Acupunctura', 377.

46   Carrubba and Bowers, 'De Acupunctura', 375.

47   Carrubba and Bowers, 'De Acupunctura', 377.

48   Carrubba and Bowers, 'De Acupunctura', 375.

49   Carrubba and Bowers, 'De Acupunctura', 375.

50   Carrubba and Bowers, 'De Acupunctura', 392.

51   Elisabeth Hsu offers two interpretations of this focus upon the material aspects of acupuncture in her article 'Outline of the History of Acupuncture in Europe', *Journal of Chinese Medicine*, 29 (1989): 28–32. First, she suggests that Ten Rhyne and his successors focused on 'static, easily observable entities' (rather than on the physiological understandings

which structured them) because they were trained in Northern Europe: 'In [seventeenth-century] medicine, two new fields of investigation were developed: physiology and microscopic anatomy.' The former, she argues, was developed and promoted in Padua, while the latter shaped medicine in the Netherlands. Consequently, 'Ten Rhyne's Dutch background suggests that explanations in terms of a more physiological outlook . . . were not of primary concern to him' (29). This argument is perhaps over-schematic, and fails to consider the internationalism of academic medicine; moreover, Ten Rhyne clearly was concerned with developing a systemic explanation for acupuncture, and not just in reporting his observations of it. Hsu's second gloss, however, is suggestive: she comments in passing, 'why should a 17th century doctor be interested in pursuing aspects of medical research which had similarities with the scholastic medicine of the Middle Ages?' (29).

52 Carrubba and Bowers, 'De Acupunctura', 391.
53 Carrubba and Bowers, 'De Acupunctura', 395.
54 Carrubba and Bowers, 'De Acupunctura', 392.
55 In contrast to the status of his writings on acupuncture, Ten Rhyne's longer account of moxibustion, although important, was only one of several contemporaneous treatises supporting the technique; its influence was thus less striking.
56 Engelbertus Kaempfer, *The History of Japan: Giving an Account of the antient and present State and Government of that Empire; of Its Temples, Palaces, Castles, and other Buildings; of Its Metals, Minerals, Trees, Plants, Animals, Birds, and Fishes; of the Chronology and Succession of the Emperors, Ecclesiastical and Secular; of the Original Descent, Religions, Customs, and Manufactures of the Natives, and of their Commerce with the Dutch and Chinese. Together with a Description of the Kingdom of Siam*, 2 vols (London, 1728). The *History* and its Appendix from the *Amoenitatum Exoticarum* were translated by J. G. Scheuchzer. Kaempfer lived from 1651–1716, and was in Japan from 1690–2.
57 Kaempfer, *History of Japan*, 'List of Subscribers'. Sloane later gave Kaempfer's materials to the British Museum, where it formed the nucleus of the East Asian collections.
58 Kaempfer, *History of Japan*, Vol. 2, 32. The entire description reads as follows: 'But now to come to the operation itself . . . The surgeon takes the needle near its point in his left hand, between the tip of the middle finger, and the nail of the forefinger, supported by the thumb, and so holds it towards the part which is to be pricked, and which must be first carefully examined, whether it be not perhaps a nerve, then with the hammer in his right hand, he gives it a knock, or two, just to thrust it through the hardish resistent [sic] outward skin. This done, he lays the hammer aside, and taking the handle of the needle between the extremities of the fore-finger and thumb, he twists it till the point runs into the body to that point, which the rules of art require, being commonly half an inch, sometimes, but seldom, an inch or upwards, in short, till it runs into the place, where the cause of the pain and distemper is supposed to be hid, where he holds it, till the patient has breathed once or twice, and then drawing it out, compresses the part with his finger, by this means, as it were, to squeeze out the vapour and spirit. The needles of the second sort are not knocked, but only twisted in.'
59 Kaempfer, *History of Japan*, Vol. 2, 34.

60 Kaempfer, *History of Japan*, Vol. 2, 63.
61 Kaempfer, *History of Japan*, Vol. 2, 29.
62 Bowers, *Medical Pioneers*, 32–3, 41. Franciscus Sylvius (Franz de la Böe) 1614–72, a Professor of Medicine at Leiden. He was an early advocate of iatrochemistry, which 'moved medicine from Galenic dogma to a recognition that health and disease were based on the chemistry of the body'.
63 Kaempfer, *History of Japan*, Vol. 2, 32, my emphasis.
64 Kaempfer, *History of Japan*, Vol. 2, 31.
65 Kaempfer, *History of Japan*, Vol. 2, 41–2.
66 Kaempfer, *History of Japan*, Vol. 2, 42.
67 'Moxa, praestantissima Cauteriorum materia, Sinensibus Japonibusque multum usitata', in Kaempfer's 1694 dissertation. Translated in Bowers and Carrubba, 'The Doctoral Thesis of Engelbert Kaempfer: "On Tropical Diseases, Oriental Medicine and Exotic Natural Phenomenon"', *Journal of the History of Medicine and Allied Sciences*, 25 (1970): 270–310.
68 Sir William Temple, *Miscellanea. By a Person of Honour* (London, 1680).
69 Hermann Busschof was a missionary; he was acquainted with Ten Rhyne in Batavia, and wrote the introductory poem in *De Arthritide* extolling moxibustion. The theories they express as to the nature of gout are almost identical, and it seems likely that Busschof derived much of his medical material directly from Ten Rhyne.
70 Again, for more information about medical and cultural responses to gout in this period, see Porter and Rousseau, *Gout*.
71 Temple, *Miscellanea*, 211–13.
72 Temple, *Miscellanea*, 211–13.
73 Temple, *Miscellanea*, 207, emphasis in the original.
74 Temple, *Miscellanea*, 211.
75 Kaempfer, *History of Japan*, Vol. 2, 39.
76 Isaac Vossius, *De Artibus et Scientiis Sinarum* (1685), quoted in Lu and Needham, *Celestial Lancets*, 286.
77 Later authors, including acupuncture's nineteenth-century advocate James Morss Churchill, scoffed at this fear, and at the idea that it was fear of the needle which drove away the pain. By then, of course, needles would have been more familiar to medical consumers.
78 Kaempfer, *History of Japan*, Vol. 2, 30.
79 Kaempfer, *History of Japan*, Vol. 2, 30.
80 Kaempfer, *History of Japan*, Vol. 2, 30.
81 Pierre Bayle, in the *Nouvelles de la République des Lettres* (Paris, 1686), 1013. Quoted in Lu and Needham, *Celestial Lancets*, 286. Michael Boym's (d.1659) *Clavis Medica ad Chinarum doctrinam de Pulsibus* was published in 1689; it is a translation of a version of the *Mo Chueh* (*Sphygmological Instructions*, approx. 940 AD). Boym's version of it described the twelve acu-tracts and Chinese ideas of circulation, but did not detail acupuncture. He did produce several quite accurate illustrations of acupoints, but described them as 'Delineatio cavitatum vel locorum pulsuum et trium partium corporis' – in other words, he interpreted them as sites by which the pulse of particular parts of the body could be taken.
82 Père J.-B. DuHalde, *A Description of The Empire of China and Chinese-Tartary, Together with the Kingdoms of Korea and Tibet: Containing the Geography and*

History (Natural as Well as Civil) of Those Countries . . . With Notes Geographical, Historical, and Critical; and Other Improvements, Particularly In The Maps, By The Translator (London, 1738–41). In this edition, the pages are mispaginated, so that there are two sets of pages sharing the numbers 229–36, but sharing nothing else. The quotation comes from the second page 235.

83 Obviously, Floyer (and a few others like him) who supported not a discrete technique but rather the use and value of pulse diagnosis did engage with the underlying theoretical structures. Floyer, however, barely mentioned particular (surgical) therapies like acupuncture or moxibustion.

84 William Wotton, *Reflections upon Ancient and Modern Learning* (London, 1694). Lu and Needham, *Celestial Lancets*, 295, discuss the paradoxical nature of Wotton's response to China.

85 Wotton, *Reflections*, 147, emphasis in original.

86 Wotton, *Reflections*, 147.

87 For discussion of Floyer, see Chapter 1.

88 Wotton, *Reflections*, 152.

89 Wotton, *Reflections*, 153.

90 See Daniel Geoffroy, *L'Acupuncture an France au XIXe siècle* (Paris, 1986); Lu and Needham, *Celestial Lancets*; for a very concise introduction to this large area, see Basil Guy 'China', in Yolton, Porter, Rogers and Stafford (eds), *The Blackwell Companion to the Enlightenment* (Oxford, 1991). During this period in France, China also acted as a stand-in for the French church and state, which were obliquely critiqued or praised through this medium. See also Basil Guy, *The French Image of China, Before and After Voltaire* (Geneva, 1963); and Huguette Cohen, 'Diderot and China', *Studies on Voltaire*, 242 (1986): 219–32. Still further, information about China often served as propaganda for the Jesuit missions there, and for their approach of accommodation; see D. E. Mungello, *Curious Land: Jesuit Accommodation and the Origins of Sinology* (Honolulu, 1989), Chapters 4 and 10.

91 Mungello, *Curious Land*, 125, Note 53, 343. I suggest that the French sought, not so much a 'culture idol', but a culture analogue, upon which to project desired changes in domestic government, etc. Of course, it is necessary to carefully distinguish between the goals of the Jesuits who selectively transmitted information about China back to Europe; the goals of their immediate editor, DuHalde; and the goals of the French reformers and radicals who employed the material provided in the *Lettres édifiantes* to a variety of ends. I would most confidently apply my interpretation to the last group. See also E. Pulleyblank and W. Beasley (eds), *Historians of China and Japan* (London, 1961) and M. G. Mason, *Western Concepts of China and the Chinese* (Cambridge, MA, 1939).

92 J.-B. DuHalde, *A Description*, Vol. 2, 183. Throughout this chapter, I have chosen to use the contemporary English translations of French works when they exist, rather than giving the quotations in French. When no contemporary English version exists, as with the *Encyclopédie Méthodique*, I include the French in the notes; translations from such works are my own. DuHalde's description appears virtually unaltered in many later works on Chinese medicine.

93 DuHalde, *A Description*, Vol. 2, 184.

94 DuHalde, *A Description*, Vol. 2, 184.

95   DuHalde, *A Description*, Vol. 2, 184.

96   Obviously, this model of a united mind and body was strikingly different from the nineteenth-century 'mind as function of body' recently discussed by Winter, Desmond and others (see Chapter 4).

97   See Anne Digby, *Making a Medical Living: Doctors and Patients in the English Market for Medicine, 1720–1911* (Cambridge, 1994), 204 and 203 respectively.

98   On the (restricted) availability of good translations, see Lu and Needham, *Celestial Lancets*, 36

99   Gerhard van Swieten, *Erläuterungen zu den Boerhaaveschen Lehrsätzen* (Vienna: 1755), quoted in Lu and Needham, *Celestial Lancets*, 293.

100 Lu and Needham argue for a fairly high level of familiarity with acupuncture in the eighteenth century; while this may be the case in continental Europe, I have found evidence only of low-level awareness of the technique in Britain. See Lu and Needham, *Celestial Lancets*.

101 Felix Vicq D'Azyr (ed.), *Médecine. Contentant, L'Hygiêne, La Pathologie, La Séméiotique & La Nosologie; La Thérapeutique Ou Matière Médicale; La Médecine Militaire; La Médecine Vétérinaire; La Médecine Légale; La Jurisprudence De La Médecine & De La Pharmacie; La Biographie Médicale. c'est-a-dire, les vies des Médecins célèbres, avec des notices de leurs ouvrages*. Tome 4, in Diderot and d'Alembert (Premiers Éditeurs), *Encyclopédie Méthodique* (Paris, 1792), 808–9.

102 Lorenz Heister, *A General System of Surgery* (London, 1743), 314.

103 Heister, *General System*, 314.

104 George Rosen, 'Lorenz Heister on Acupuncture: an Eighteenth Century View', *The Journal of the History of Medicine and the Allied Sciences*, 30 (1975): 386–8, at p. 388, briefly analyses Heister's description of acupuncture, and Heister's significance as an authority. Noting the longevity of his works and his position as 'one of leading surgeon of the period', Rosen concludes that Heister 'can be regarded as an accurate contemporary reflection of medical opinions on this procedure'.

105 F. Dujardin, *Histoire de la Chirurgie, Depuis son Origine jusqu'à nos Jours*, Tome 1 (Paris, 1774), 77. 'Comme ils n'ont point de physique, presque aucune conoissance des parties du corps humaine & leurs usages, ni par conséquent des causes des maladies, leur Médecine, dénuées de tout principe, n'est qu'un amas informe de systèmes, de tâtonnments, de conjectures.'

106 Dujardin, *Histoire de la Chirurgie*, 77, 'le plus souvent l'ouvrage de l'imagination; ainsi, toutes les conoissances qu'ils en déduisent, ne sauroient être forts solides'.

107 Dujardin, *Histoire de la Chirurgie*, 83, 'paroîtra ridicule & pitoyable; cependant à travers le brouillard, il perce quelquefois de légères lueurs de vraisemblance'.

108 Dujardin, *Histoire de la Chirurgie*: 83, 'by means of the nerves, veins and arteries'.

109 Dujardin, *Histoire de la Chirurgie*, 85, 'Malgré toutes les hypothèses qui défigurent cet empyrisme, l'expérience a quelquefois servi les Practitiens de la Chine.'

110 That this shift had not yet taken place in Britain is evident from Gillan and Staunton's reactions to Chinese practice some years later (see Chapter 1).

111 Dujardin, *Histoire de la Chirurgie*, 88, '[I]ls ont deux autres remèdes qu'ils empruntent de la Chirugie, & qu'ils regardent comme spécifiques. Toute

maladies qui résiste à ceux-ci, qui sont le moxa & la ponction avec les éguilles, est reputé incurable.'

112 Dujardin, *Histoire de la Chirurgie*, 88, 'à peu pres de la même manière qu'en Europe on a recours à la saignée & à la purgation, pour diminuer la pléthore ou prévenir l'orgasme des humeurs'.

113 Dujardin, *Histoire de la Chirurgie*, 90, 'ce remède . . . jete les malades dans les angoisses qui vont jusqu'à la syncope, quand on porte l'appliquation à un certain excès'.

114 Dujardin, *Histoire de la Chirurgie*, 91, 'des figures singulières'.

115 Dujardin, *Histoire de la Chirurgie*, 91, '[O]n y voit la marche des vaisseaux, telle qu'ils imaginent. Les endroits qu'il faut piquer, sont désignés par des points verts, & ceux qu'on doit brûler, par des points rouges. La connoisance de ces endroits a paru si importante, qu'ayant été depuis érigée en art, elle est excercée par des espèce d'Experts comme sont chez nous les Bandagistes, & c'.

116 Dujardin, *Histoire de la Chirurgie*, 91–2, 'Les lieux de l'application diffèrent, selon le genre des maladies, le caractère des humeurs & la nature des parties subjacentes. Les préceptes de l'art tiennent à la distribution des vaisseaux & aux mouvement du sang, que les Chinois & les Japonois connoisent mieux, à ce que prétend Ten Rhyne, qu'aucune nation de l'Europe.'

117 He argues that the French brought 'l'angeiologie & les autres parties de l'Anatomie à la démonstration la plus complete'. Dujardin, *Histoire de la Chirurgie*, 91.

118 Dujardin, *Histoire de la Chirurgie*, 91–2, 'Il ne faut pas croire qu'une légère erreur dans le local précis, fût un obstacle au succès du remède; cependant plusieurs faits prouvent qu'il importe de ne point s'écarte des principes . . . Ce qu'on peut dire de plus certain, c'est que, dénués d'anatomie comme ils sont, ils ne peuvent tenir les principes qu'ils se sont faits dans l'application du moxa & des aiguilles, que d'un nombre infini d'expériences qu'ils multiplient sans cesse' (91).

119 As was discussed above, Ten Rhyne particularly stressed the idea that anatomical knowledge, combined with experiment and experience, could guide Europeans in performing acupuncture if they chose not to follow the Chinese maps. Of course, this reading does not reflect East Asian practice or understandings of the maps' function.

120 Dujardin, *Histoire de la Chirurgie*, 98, 97, 'la partie malade' and 'la partie où le mal a pris naissance'.

121 Dujardin, *Histoire de la Chirurgie*, 91–2, 'vents qui se glissent entre le perioste & les os: fait dont il prètend s'être assuré par l'observation.'

122 Dujardin, *Histoire de la Chirurgie*, 91–2, 'Un malheur attaché à l'humanité, . . . qui s'oppose au progrès de nos connoissances, c'est que les Observateurs, même de bonne foi, mais prévenus, rapportent tout ce qu'ils voient à l'idée qui les occupe. Cette idée favorite est un enfant gâté, que l'imagination pare toujours aux dépens de la raison & de la vérité'.

123 Dujardin, *Histoire de la Chirurgie*, 94, '[N]otre Médecine est devenues trop discoureuse; c'est que chez nous l'étude des parties a fait négliger la science pratique de l'ensemble, ou de cette conspiration des parties entr'elles, si bien observée par Hippocrate & par tous les vrais Médecins: en cela seul, la

Médecine des Chinois, toute empyrique, toute imparfaite qu'elle est, même à cet égarde, est digne de quelque attention.'

124 Dujardin, *Histoire de la Chirurgie*, 98, 'La ponction... n'agit vraisemblablement qu'en appelant dans la partie irritée une plus grande affluence d'humeurs, à moins que l'imagination, dispensatrice de tant biens & de maux physiques & moraux, n'aide l'action de ce remède'.

125 R. James, *A Medicinal Dictionary; Including Physic, Surgery, Anatomy, Chymistry, and Botany, in all their Branches Relative to Medicine... and an Introductory Preface, tracing the Progress of Physic, and Explaining the Theories which have Principally Prevail'd in All the Ages of the World*, Vol. 1 (London, 1743), Pt. I, 'acupuncture' (pages unnumbered).

126 James, *A Medicinal Dictionary*, 'acupuncture'.

127 *The Modern Part of the Universal History*, Vol. 3 (London, 1759), 649. Intriguingly, the *Universal History* does mention the existence of specific points proper to the operation of acupuncture, in the context of a long story about the Emperor Kang Hsi's desire for a translation of western anatomical texts. Upon receiving such a translation, 'that prince, recollecting that he had seen, among other of his rarities, a statue of about three feet high, cast in copper, on which were, as he imagined, all the veins and arteries, delineated in their proper places... To their great surprize, they found those lines all parallel to each other, and almost all of the same length, without any the least resemblance either to veins or arteries, or answering to their true situation or number... [T]hey soon found that those lines were traced on the figure with no other view than to point out the place that were proper to let blood by, by the operation lately mentioned, called acupuncture, or by the help of coarse needles, in cases of rheumatism, gout, sciatica, & c.' (654).

128 Vicq D'Azyr (ed.), *Médecine*, Tome 4, 808–9. This quotation indicates some of the ambivalence with which China and things Chinese were regarded by the close of the eighteenth century, with its urgings that China should be imitated, despite its flaws; similarly, its medicine was dangerous, and exceptional at the same time.

129 For a brief and broad overview of this process, see Russell Maulitz, 'The Pathological Tradition', in Porter and Bynum (eds), *The Routledge Companion to the History of Medicine*, Vol. 1 (New York, 1995), 169–91; for a more tightly focused look at the period covered here, see Russell C. Maulitz, *Morbid Appearances: the Anatomy of Pathology in the Early Nineteenth Century* (Cambridge, 1987); J. E. Lesch, *Science and Medicine in France: the Emergence of Experimental Pathology, 1790–1855* (Cambridge, MA, 1984) tells the story from a somewhat different perspective.

130 For a detailed examination of late eighteenth- and nineteenth-century nervous models, and their close relationship with galvanism and electricity, see Edwin Clarke and L. S. Jacyna, *Nineteenth-Century Origins of Neuroscientific Concepts* (Berkeley, 1987), especially Chapter 5. The authors focus on Germany and romantic biology, but also discuss French and British medical science. One drawback to their work is that in documenting the changing notions of the nerve and of nervous activity, they look at the leading edge, without observing its vast distance from the lagging edge. Thus they discuss the decline of ideas of nervous fluid in the late eighteenth century, while

medical practitioners in this study were citing 'nervous fluid' well into the 1850s and beyond.

131 See Maulitz, *Morbid Appearances*; Christopher Lawrence, 'Democratic, Divine and Heroic: the History and Historiography of Surgery', in C. Lawrence (ed.), *Medical Theory, Surgical Practice: Studies in the History of Surgery* (London, 1992), 1–47; Malcolm Nicholson, 'Giovanni Battista Morgagni and Eighteenth-Century Physical Examination', in Lawrence, *Medical Theory, Surgical Practice*, 101–34. Although the union of medicine and surgery in revolutionary and immediately post-revolutionary France is not the major theme of Nicholson's article, he deals with the subject concisely and clearly: 'The cognitive consequence of this union ... was a body of medical knowledge in which internal disease was newly conceived in localised, structural anatomic terms, as opposed to the whole-body humoral pathology of eighteenth-century physic' (122). The 'intellectual invasion of the body by surgeons' (to use Lawrence's phrase) was paralleled by a manual exploration of the body by physicians.

132 Robert Darnton, *Mesmerism and the End of the Enlightenment in France* (Cambridge, MA, 1968); Daniel Geoffroy, *L'Acupuncture an France au XIXᵉ siècle*, (Paris, 1986); Alison Winter, *Mesmerized: Powers of Mind in Victorian Britain* (Chicago, 1998). See also Chapter 4.

133 De la Roche and Petit-Radel (eds), *Chirurgie*, in Diderot & d'Alembert, *Encyclopédie Méthodique*, Tome 1, 59, 'Les Nations dont nous parlons, quoique d'ailleurs très-industrieuses et très-sensées, exécutent cette étrange opération, non-seulement à la tête, mais encore aux bras, aux jambes, et à plusieurs autre parties; ils vont même jusqu'à percer le ventre des femmes enceintes.'

134 De la Roche and Petit-Radel, *Chirurgie*, 59, 'Comme cette opération n'est practiquée nulle part en Europe, ne nous y arrêterons pas davantage'.

135 Vicq D'Azyr, *Médecine*, Tome 2, 185, 'Dans toutes ces maladies, on perce, dit-on, l'endroit même où est le siége du mal, ou celui dans lequel le mal a pris naissance.'

136 Vicq D'Azyr, *Médecine*, Tome 2, 185, 'L'expérience a appris aux peuples de l'Orient que ... des ponctions multiplées, & plus ou moins profondes, faites avec des aiguilles ... deviennnent un secours très-efficace, & que souvent les douleurs les plus aiguës s'appaisent aussi-tôt après qu'on a fait cette opération.'

137 Vicq D'Azyr, *Médecine*, Tome 2, 186–7, 'Les chinois ... pensent que le principe de la plupart des maladies consiste dans des vapeurs nuisibles renfermées dans les parties souffrantes ... & dont il n'est besoin, pour guérir, que de les delivrer. C'est, suivant le systême adopté par ces peuples, ce que produisent l'acupuncture, en ouvrant à ces vapeur mal-faisantes des issues favorables, & le moxa, en les attirant à la surface du corps, & en les y consumant.'

138 Vicq D'Azyr, *Médecine*, Tome 2, 188, 'acupuncture est un procédé que l'on doit ranger parmi les moyens irritans & stimulans; ... elle peut ainsi compter des spasmes violens, & rétablir la sensibilité in les organes où cette fonction a été affoiblie'.

139 All Vicq D'Azyr, *Médecine*, Tome 2, 188, 'comme des remèdes fameux dans les autres pays'; 'ces prétendues humeurs mal-faisantes'; 'ceux qui conois-

sent bien l'économie animale, & qui ont profondément médité sur la nature des maladies'.

140 Vicq D'Azyr, *Médecine*, Tome 2, 188, 'Toujours est-il certain que ces effets jettent un grand jour sur plusieurs questions des plus importantes dans l'art de guérir'.

141 L. V. J. Berlioz, *Mémoires sur les Maladies Chroniques, les Évacuations Sanguines et l'Acupuncture* (Paris, 1816), 298, 'Les éloges donnés à l'acupuncture par Kempfer et Then-Ryne [sic] sont justes et mérités. On a lieu d'être étonné que, depuis un siècle et plus que ce moyen curatif est connu en Europe, aucun médecin ne l'ait essayé jusqu'ici.' This translation of Berlioz's words appears in James Morss Churchill, *A Treatise on Acupuncturation* (London, 1821), 25.

142 Berlioz, *Mémoires*, 298, 'l'acupuncture, en dissipant les accidens, démontre que le désordre du système nerveux leur avait donné naissance'.

143 Berlioz, *Mémoires*, 298, 'Les affections nerveuses simples démontrent spécialement combien l'acupuncture mérite l'attention des médecins; car il est peu de remède qui jouissent d'une activité aussi prompte, et qui produisent des effets aussi mervielleux'.

144 Berlioz, *Mémoires*, 296–7. For reasons of space, I have omitted the lengthy original.

145 Berlioz, *Mémoires*, 301. In his second case study, having seen from an accident in the first that needles penetrating the epigastric region did no apparent harm to his patient, he inserted the needles so deeply that he believed he had pierced the stomach. 'Cette opération a été accusée de témérité par les membres de la Société de Médecine de Paris, composant la commission nommée pour faire le rapport sur les ouvrages envoyés au concourse de 1811.'

146 Berlioz, *Mémoires*, 118, 'La correspondance des masses du tissu cellulaire n'est pas non plus à négliger dans le traitement des maladies chroniques.'

147 Berlioz, *Mémoires*, 105, 'l'excitation de la peau stimule par sympathie les membranes muqueuses; mais telles ou telles région de l'enveloppe cutanée ont plus de rapport avec telle ou telle autre région tapissée par les membranes muqueueses'.

148 Berlioz cites the *Zoonomia* as the source of this curious piece of information; it does indicate that the doctrine of local sympathy was fairly widespread, although one wonders whether this particular example came from traditional healing rather than academic medicine – it has a certain earthy particularity to it often lacking in the more 'elevated' schools of medicine.

149 A. Carraro, 'Acupuncture of the Heart in Apparent Death', *Provincial Medical and Surgical Journal* (15 May 1841): 140.

150 Berlioz, *Mémoires*, 310–11, 'ce qui porte à croire que l'acupuncture n'agit point en dètruisant une irritation par une autre; ... je le répète, elle n'a jamais plus de succès que lorsqu'elle est peu ou point douloureuse. Il paraît, au contraire, que ce remède agit en stimulant les nerfs, ou en leur restituant un principe dont ils étaient privés par l'effet de la douleur... Vraisemblablement la communication du choc galvanique produit par un appareil de Volta, accroîtrait les effets médicaux de l'acupuncture.'

151 For more on the subsequent history of acupuncture in France, see Geoffroy, *L'Acupuncture*; Roger Baptiste, *L'Acupuncture et son Histoire: Avantages et*

Inconvénients d'une Thérapeutique Millénaire (Paris 1962), 25–49; J. Bossy, 'The History of Acupuncture in the West', in Teizo Ogawa (ed.), *History of Traditional Medicine: Proceedings of the 1st and 2nd International Symposia on the Comparative History of Medicine – East and West* (Tokyo, 1986), 363–400.

3 Sharpening the Needle

1 Anon, *The Modern Part of the Universal History*, Vol. 4 (London, 1759), 647.
2 Erasmus Darwin actually used the term 'acupuncture' in 1794. In Darwin, *Zoonomia; or the laws of organic life*, 3rd edn, Vol. 3 (London, 1801), 254, he asked: 'In cases of strangulated hernia, could acupuncture, or puncture with a capillary trocar be used with safety and advantage to give exit to air contained in the strangulated bowel? Or to stimulate it into action?' His use of the term suggests that he was merely using Latin shorthand, and not actually refering to acupuncture as it emerged from China and Japan. Acupuncture is not mentioned elsewhere in his work. It is possible that Coley might have come across this reference, but I think it more likely that his information came from the sources he cited himself.
3 Coley, 'A Case of Tympanites, in an Infant, relieved by the Operation of the Paracentesis. With Remarks on the Case; and a Critical Analysis of the Sentiments of the Principal Authors who have written on the Disease. To which is subjoined an Account of the Operation of the Acupuncture, as Practised by the Japanese in the Diseases analogous to the Tympany', *The Medical and Physical Journal*, 7 (1802): 235–8.
4 Coley, 'A Case', 235.
5 Lorenz Heister, *A General System of Surgery in Three Parts* (London, 1743), 314; and *Universal History*, Vol. 4, 18.
6 *Universal History*, Vol. 3, 599, my emphasis.
7 *Universal History*, Vol. 3: 647, my emphasis.
8 For more on British attitudes towards dissection, see Coral Lansbury, *The Old Brown Dog: Women, Workers, and Vivisection in Edwardian England* (Madison, 1985), especially Chapter 3; Ludmilla Jordanova, *Sexual Visions: Images of Gender in Science and Medicine Between the 18th and 20th Centuries* (Madison, 1989); and Ruth Richardson, *Death, Dissection and the Destitute: a Political History of the Human Corpse* (London, 1987).
9 Coley, 'A Case', 235.
10 Coley, 'A Case', 235.
11 Coley, 'A Case', 235.
12 See Edwin Clarke and L. S. Jacyna, *Nineteenth-Century Origins of Neuroscientific Concepts* (Berkeley, 1987) for more on controversies surrounding 'nervous fluid' and electrical explanations. Remember also that this was a time when in Britain excess rationalism in philosophy and science was clearly seen to lead to bloody political madness.
13 In fact, the *Universal History* excerpt itself offers a perfect example of whence this understanding of acupuncture's function and functionality sprang; as the author describes the application of acupuncture in Japan, he mentions the practice of compressing the site of puncture, 'in order to force the morbific vapour or spirit out' (Coley, 'A Case', 237). As Ten Rhyne noted

in his discussion of the type of needles in the *Dissertatio de Arthritide*, the use of puncturing for certain types of fluid retention had been practised in Europe prior to the introduction of acupuncture either as a technique or as a term. Apparently, given the claims of novelty made in the nineteenth century for this operation as a form of 'acupuncture', the needle had been entirely replaced by the lancet at some point in the eighteenth century.

14 Coley 'A Case', 237–8. The 'others' referred to in this quotation are Chinese and Japanese practitioners, not innovative western counterparts.

15 Sir William Temple's exhortations on the subject took some time to filter into the medical marketplace, perhaps because of their initially limited elite audience.

16 Roy Porter, 'The Rise of Medical Journalism in Britain to 1800', in W. F. Bynum, Steven Lock and Roy Porter (eds), *Medical Journals and Medical Knowledge: Historical Essays* (London, 1992), 6–28, at p. 18. This collection of essays offers a variety of perspectives, quantitative and qualitative, on medical journalism in late eighteenth- and early nineteenth-century Britain. *The Medical and Physical Journal* was owned by Richard Phillips, whose sympathies were republican and radical, and who pursued medical reform zealously. In 1815, an 'Address' from 'the editors' (presumably Samuel Fothergill, listed as editor on the masthead) noted that, 'communications with the name of the author will always claim a priority, and that even these will be distinguished as the subject may be of a temporary or transient nature. By attention to the latter, their Journal has now become a register for events which are partly forgotten, by fresh ones which arise, more interesting for their novelty.' 'Address', *The Medical and Physical Journal*, 33 (1815): 1.

17 Jean Loudon and Irvine Loudon, 'Medicine, Politics and the Medical Periodical, 1800–1850', in Bynum, Lock and Porter, *Medical Journals and Medical Knowledge*, 49–69. They credit *The Medical and Physical Journal* with a 'cool analytical approach to medical politics . . . head and shoulders above most anti-establishment journals', at p. 60.

18 'Medical and Philosophical Intelligence. "State of Medicine in China; by M. Page, of Orleans." ' *The London Medical and Physical Journal*, 33 (1815): 247–8.

19 Hugh Murray, John Crawfurd, Peter Gordon, Captain Thomas Lynn, William Wallace and Gilbert Burnnet, *An Historical and Descriptive Account of China; Its Ancient and Modern History, Language, Literature, Religion, Government, Industry, Manners, and Social State; Intercourse with Europe from the Earliest Ages; Missions and Embassies to the Imperial Court; British and Foreign Commerce; Directions to Navigators; State of Mathematics and Astronomy; Survey of its Geography, Geology, Botany, and Zoology*, Vol. 3 (Edinburgh, 1836), 283–4.

20 John Francis Davis, *The Chinese: a General Description of the Empire of China and Its Inhabitants* (London, 1837), 73. Davis's account was based on a twenty-year residence beginning with the 1816 Lord Amherst Embassy to Peking, by the end of which he was serving as His Majesty's Chief Superintendent in China.

21 Clarke Abel, *Narrative of a Journey in the Interior of China, and of A Voyage to and from that Country, in the Years 1816 and 1817; Containing an Account of the Most Interesting Transactions of Lord Amherst's Embassy to the Court of Pekin, and Observations on the Countries which it Visited* (London, 1818), 216–17, my

emphasis. Abel was a Fellow of the London Society and of the Geological Society, Chief Medical Officer and Naturalist to the Embassy.

22 Abel, *Narrative of a Journey*, 216–17.

23 Abel, *Narrative of a Journey*, 218.

24 British responses to the Chinese people were not monolithic, but in general commentators were far more favourably disposed towards the Chinese population and manufactures than towards the government, the Manchu elite and Chinese technological and scientific productions.

25 Henry Ellis, *Journal of the Proceedings of the Late Embassy to China; comprising a correct narrative of the public transactions of the Embassy, of the voyage to and from China, and of the journey from the mouth of the Pei-ho to the return to Canton. Interspersed with observations upon the face of the country, the polity, moral character, and manners of the Chinese Nation* (London, 1817), 40.

26 Ellis, *Proceedings of the Late Embassy*, 489.

27 See R. Darnton, *Mesmerism and the End of the Enlightenment in France* (Cambridge, MA, 1968); Alison Winter, *Mesmerized: Powers of Mind in Victorian Britain* (Chicago, 1998) and below, Chapter 4.

28 Murray et al., *An Historical and Descriptive Account of China*, Vol. 2, 85.

29 Murray et al., *An Historical and Descriptive Account of China*, Vol. 2, 85.

30 Murray et al., *An Historical and Descriptive Account of China*, Vol. 2, 86.

31 Murray et. al., *An Historical and Descriptive Account of China*, Vol. 2, 87–8.

32 'Review. Medical and Surgical Cases: Selected During a Practice of Thirty-Nine Years. – By Edward Sutleffe', *Lancet* (April 22, 1826): 102–9, at p. 106.

33 Edward Sutleffe, *Medical and Surgical Cases; Selected During a Practice of Thirty-Eight Years*, Vol. 1 (London, 1824–5), 45, 272–3.

34 'Medical and Surgical Cases', *Lancet*, 107.

35 For a fascinating treatment of continuities and change in Chinese medicine, see Paul Unschuld, *Medicine in China: a History of Ideas* (Berkeley, 1985). See also Ralph Dale and Yan Cheng, 'An Outline History of Chinese Acupuncture: the Main Developments, Contributors and Publications', *American Journal of Acupuncture*, 21 (1993)· 355–93; this article does exactly what its title suggests, offering at least limited access to sources otherwise unavailable in English.

36 'Critical Analysis of English and Foreign Literature Relative to the Various Branches of Medical Science. Division II', *The London Medical and Physical Journal*, 48 (1822): 518, emphasis in original. The editors at this point were Roderick Macleod and John Bacot. It is likely, considering his future involvement with the subject of Chinese medicine, that Macleod wrote this review.

37 'Retrospective of Foreign Medical Science and Literature, for the year 1819: I. Succinct Analysis of Foreign Periodical Literature', *London Medical Repository*, 13 (1820): 33–87. This section was extracted from *Bulletins de la Faculté de Médecine de Paris, et de la Société établie dans son sein* (February 1819): 38.

38 In this case, the tendency to translate Chinese theoretical terms quite physically and literally may have served the meaning of the Chinese text better than it usually did. The phrase 'congestions of the head and abdomen' approaches a physical description of the Chinese model of diseases for which acupuncture is useful – those in which the natural flow of the *qi*

(vital energy) has become blocked, creating pools of misplaced energy – congestions of *qi*.

39   'Retrospective of Foreign Medical Science and Literature', 3.

40   For the classic account see Stanley J. Reiser, *Medicine and the Reign of Technology* (Cambridge, 1978), and also Lindsay Granshaw, ' "Upon This Principle I Have Based A Practice": the Development and Reception of Antisepsis in Britain, 1890–1914', in John Pickstone (ed.), *Medical Innovations in Historical Perspective* (New York, 1992), 17–46; Malcolm Nicholson, 'The Introduction of Percussion and Stethoscopy to Early Nineteenth-Century Edinburgh', in W. F. Bynum and Roy Porter (eds), *Medicine and the Five Senses* (Cambridge, 1993), 134–53; and Chris Lawrence, 'Incommunicable Knowledge: Science, Technology and the Clinical Art in Britain, 1850–1914', *Journal of Contemporary History*, 20 (1985): 503–20.

41   James Morss Churchill, *A Treatise on Acupuncturation; Being a Description of a Surgical Operation Originally Peculiar to the Japonese and Chinese, and by them denominated Zin-King, Now Introduced into European Practice, with Directions for its Performance and Cases Illustrating its Success* (London, 1822).

42   Churchill, *Treatise*, 4.

43   Churchill, *Treatise*, 5.

44   Churchill, *Treatise*, 10.

45   Churchill, *Treatise*, 10.

46   Churchill, *Treatise*, 10.

47   Churchill, *Treatise*, 11–12, my emphasis.

48   Churchill, *Treatise*, 13.

49   Churchill, *Treatise*, 13.

50   Churchill, *Treatise*, 13–14.

51   Remember that Coley's introduction of his piece was 'It may yet perhaps be thought by the English Surgeons on some occasions to be worth imitating; and as the method of doing so is both curious and but little known, the following detail of it, from the volume mentioned, is transcribed for more public information, and to conclude the subject . . .', Coley, 'A Case', 237.

52   Churchill, *Treatise*, 22–3, my emphasis.

53   See David Arnold, *Colonizing the Body: State Medicine and Epidemic Disease in 19th Century India* (Berkeley, 1993), 43–58. For a more extended, if problematic, treatment of this period, see David Kopf, *British Orientalism and the Bengal Renaissance: the Dynamics of Indian Modernization 1773–1835* (Berkeley, 1969). As discussed elsewhere in this volume, trends in Orientalist scholarship have consistently influenced western perceptions and representations of acupuncture.

54   Churchill, *Treatise*, 12.

55   Churchill, *Treatise*, 23–4, my emphasis. In Chapter 4, I will discuss the nature of Churchill's practice, and the implications of his emphasis on experiment for his patients – particularly for 'labouring persons'.

56   In fact, this operation preceded the arrival of the Asian form of acupuncture in Europe. Ten Rhyne described puncturing in cases of dropsy in his introductory disquisition on needles. However, it seems to have died out in the intervening years, as nineteenth-century physicians claimed their use of 'acupuncture' in dropsy was novel and innovative.

57   Churchill, *Treatise*, 25.

58 Churchill, *Treatise*, 45.
59 Churchill, *Treatise*, 71. In fact, this is the same sensation described by modern acupuncturists as following the correct stimulation of an acupuncture point.
60 Churchill, *Treatise*, 71–2, my emphasis.
61 Churchill *Treatise*, 85
62 As Stephen Jacyna notes in ' "Mr Scott's Case": a View of London Medicine in 1825', historians of medicine can legitimately consider the 1820s as either the end of the long eighteenth century or as 'the outset of a new era in the history of clinical medicine'. Contained in Roy Porter (ed.), *The Popularization of Medicine 1650–1850* (London, 1992), 252–86, at p. 254.
63 See Darnton, *Mesmerism*.
64 This technique has been used again by 'medical acupuncturists' in the present century. See below, Chapter 5.
65 Compare Churchill, *Treatise*, 22–3, with 'Retrospective of Foreign Medical Science and Literature', *London Medical Repository*, as discussed above.
66 The question of the extent and depth of acupuncture's diffusion and popularity in Britain will be discussed in Chapter 4.
67 For a useful overview see Chris Lawrence, 'Democratic, Divine and Heroic: the History and Historiography of Surgery', in Lawrence (ed.), *Medical Theory, Surgical Practice: Studies in the History of Surgery* (London, 1992), 1–47; see also Winter, *Mesmerized*, 166–9, on attitudes towards pain, and the use of pain as a tool in the medical campaign against surgeons' call for equal status.
68 Intriguingly, the American response to Churchill's book, published in the US half a decade later, was substantially more hostile, and described the tone as, at best, 'the tone of youth' and, at worst, as arrogant and boastful. See Dorothy Rosenberg, 'Acupuncture and U.S. Medicine: a Socio-Historical Study of the Response to the Availability of Knowledge' (PhD dissertation, University of Pittsburgh, 1977).
69 Adrian Desmond, *The Politics of Evolution: Morphology, Medicine, and Reform in Radical London* (Chicago, 1992). In 'Democratic, Divine and Heroic', Christopher Lawrence notes 'During the years of the French wars, British surgeons gradually changed the tone of their histories and began to introduce new themes . . . British surgeons . . . increasingly cited France as having had a more distinguished history of surgery' (p. 6).
70 Of course, using these articles in such a way does require some care. They are generally anonymous, presumably produced by the editors of the journals in question, but not necessarily so.
71 'Analytical Review, IV: *A Treatise on Acupuncturation: Being a Description of a Surgical Operation, originally peculiar to the Japonese* [sic] *and Chinese, now introduced into European Practice; with Directions for its Performance, and Cases illustrating its Success*. By James Morss Churchill, M.R.C.S., London, 1821', *London Medical Repository*, 17 (1822): 236–7, at p. 236.
72 'Analytical Review, IV', *London Medical Repository*, 236.
73 'Analytical Review, IV', *London Medical Repository*, 236.
74 'Analytical Review, IV', *London Medical Repository*, 237.
75 See Winter, *Mesmerized* for an enlightening presentation of the changing status of subjective experience as a source of authority in experiment and medicine during and after this period. In traditional Chinese acupuncture,

patient sensations do play a role in finding the precise location of acu-points, but without the assistance of a map of the channels and a sense of the relationship between different points, only the patient's unique electrical-sensation would identify an active point. Jukes's letter and this review are unusual in reporting this sensation – perhaps Jukes was simply lucky in his insertion, but he was certainly a close observer.

76 'Analytical Review, IV', *London Medical Repository*, 237.
77 'Acupuncturation', *Lancet* (9 November 1823): 200–1, at p. 200.
78 'Acupuncturation', *Lancet*, 201. Of course, a contemporary reader would probably have interpreted this statement in terms of the familiar anatomical models of the day, rather than a surface map of specific points.
79 'Acupuncturation', *Lancet*, 201, my emphasis.
80 'An Historical Essay on the State of the Medical Sciences During the Last Six Months', *The London Medical and Physical Journal*, 50 (1823): 1–66, at p. 57.
81 'An Historical Essay', 57.
82 Scotus, 'Sciatica treated by Acupuncture, with Dr Alison's opinion on the mode of its Operation', *Lancet* (19 May 1827): 190–1, at p. 190.
83 Alison Winter's recent work on mesmerism explores at length the pains taken by British medical reformers to strip authority from patient testimony and self-reporting, vesting it instead in professional readings of the patient's physical signs. The British response to acupuncture confirms this trend, and establishes a slightly earlier starting point for Wakley's efforts in this direction. See Winter, *Mesmerized*, Chapter 4.
84 Scotus, 'Sciatica', 190.
85 Scotus, 'Sciatica', 190.
86 Scotus, 'Sciatica', 190.
87 Scotus, 'Sciatica', 191.
88 Scotus, 'Sciatica', 191.
89 Scotus, 'Sciatica', 191.
90 Scotus, 'Sciatica', 191. It is worth noting that these objections to acupuncture rehearsed the medical objections to mesmerism which were so soon to follow. See below.
91 'Art. XIV. [Review of recent work on Acupuncture]', *The Edinburgh Medical and Surgical Journal*, 27 (1827): 190–200, at p. 191.
92 'Art. XIV', *Edinburgh Medical and Surgical Journal*, 191.
93 'Art. XIV', *Edinburgh Medical and Surgical Journal*, 191–2.
94 'Art. XIV', *Edinburgh Medical and Surgical Journal*, 192.
95 'Art. XIV', *Edinburgh Medical and Surgical Journal*. The author's rather hesitant phrasing was that 'a little precise information on the relative success of treatment under these and other analogous circumstances, might go far to determine what truth there is in the conjecture, to which one is naturally apt to be led, and which (strange as it may seem) has hardly been started, and not at all investigated in any of these treatises before us – that acupuncture belongs to the class of remedies which acts through the medium of emotions of the mind', 197.
96 'Art. XIV. [Review of recent work on Acupuncture, continued.]', *The Edinburgh Medical and Surgical Journal*, 27 (1827): 334–49, at p. 334.
97 'Art. XIV Cont.', *Edinburgh Medical and Surgical Journal*, 193.
98 'Art. XIV Cont.', *Edinburgh Medical and Surgical Journal* (1827), 193.

99   'Art. XIV Cont.', *Edinburgh Medical and Surgical Journal* (1827), 199–200.
100   'Art. XIV Cont.', *Edinburgh Medical and Surgical Journal*, 335.
101   'Art. XIV Cont.', *Edinburgh Medical and Surgical Journal*, 337.
102   'Art. XIV Cont.', *Edinburgh Medical and Surgical Journal*, 337.
103   'Art. XIV Cont.', *Edinburgh Medical and Surgical Journal*, 337.
104   'Art. XIV Cont.', *Edinburgh Medical and Surgical Journal*, 338.
105   'Art. XIV Cont.', *Edinburgh Medical and Surgical Journal*, 339
106   'Art. XIV Cont.', *Edinburgh Medical and Surgical Journal*, 349.
107   'Medical and Physical Intelligence. Physiology. 2. On the Galvanic Phenom-
      ena which accompany the Acupuncturation', *London Medical and Physical
      Journal*, 53 (1825): 434.
108   'Medical and Physical Intelligence', 434, my emphasis.
109   'Medical and Physical Intelligence', 434. These results were taken from
      Bulletin des Sciences Médicales (March 1825).
110   'Pelletan on the Galvanic Phenomena of Acupuncturation', *Anderson's Quar-
      terly Journal of the Medical Sciences*, 2 (1825): 428.
111   'Foreign Department. Analysis of Foreign Medical Journals. Revue Medicale
      – June 1825. *On the Electro-Magnetic Phenomena Observed in Acupuncture*. By
      M. Pouillet', *Lancet* (6 August, 1825): 152–3, at p. 152.
112   Just as a fingernail sketch, of thirty-nine articles published in medical
      journals between 1802 and 1831, including both foreign and domestic
      case studies, only eight discuss its mode of action in any detail. Less than
      five additional cases discuss the MO to the extent of dismissing one or
      another explanation of it.
113   Frederic Finch, 'Case of Anascara in which Acupuncturation was Success-
      fully Employed, and the Fluid Discharged by it', *London Medical Repository*,
      19 (1823): 205–6.
114   Frederic Finch, 'Case of Trismus, &c., approaching to Tetanus, supervening
      to a lacerated Wound, successfully treated by Acupuncturation', *London
      Medical Repository*, 19 (1823): 403–4, at p. 403.
115   T. W. Wansbrough, 'Acupuncturation', *Lancet* (30 September 1826): 846.
116   Wansbrough, 'Acupuncturation', *Lancet*, 846.
117   Finch, 'Case of Anascara', 205–6.
118   Harry William Carter, 'A General Report of the Medical Diseases treated in
      the Kent and Canterbury Hospital, from January 1st to July 1st, 1823, with a
      particular Account of the more important Cases', *London Medical Repository*,
      19 (1823): 386–402, at p. 398; continued as 'Case No. XVIII "On Acupunc-
      tura"', *London Medical Repository*, 19 (1823): 455–6, at p. 456.
119   Finch, 'Case of Trismus', 403.
120   'I shall not here hazard an hypothesis of the modus operandi of acupunc-
      turation on the animal oeconomy; but at the same time I am free to confess
      myself sceptical on the creed that its effects by the *escape* of *air* from the
      cellular membrane through the punctures made by the needles.' Wans-
      brough, 'Acupuncturation', 848.
121   Wansbrough, 'Acupuncturation', 848. It will suffice to point out that this is
      an even more simplistic form of the mis-translated Chinese theory of acu-
      puncture's mode of action. Whether or not Wansbrough was aware of the
      origins of this theory is unclear, but clearly the theory was circulating in
      some way.

122 Wansbrough, 'Acupuncturation', 848.
123 James Morss Churchill, *Cases Illustrative of the Immediate Effects of Acupuncturation, in Rheumatism, Lumbago, Sciatica, Anomalous Muscular Diseases, And in Dropsy of the Cellular Tissue; Selected from various sources, and intended as an Appendix to the Author's Treatise on the Subject* (London, 1828), 3.
124 W. F. Bynum, S. Lock and R. Porter (eds), *Medical Journals and Medical Knowledge: Historical Essays* (London, 1992). I address the subject of formal and informal medical education in Chapter 4.
125 John Elliotson, 'Acupuncture', in John Forbes, Alexander Tweedie and John Conolly (eds), *Cyclopaedia of Practical Medicine; Comprising Treatises on the nature and treatment of diseases, Materia Medica and Therapeutics, Medical Jurisprudence, etc. etc.*, Vol. 1 (London, 1833), 32–3.
126 Elliotson, 'Acupuncture', 32.
127 Elliotson, 'Acupuncture', 32–3. This description, slightly edited, also formed part of the Appendix to one of several published collections of Elliotson's lectures. John Elliotson, *The Principles and Practice of Medicine; Founded on the most extensive experience in public hospitals and private practice; and as developed in a course of Lectures, delivered at University College, London*, with notes and illustrations by Nathaniel Rogers (London, 1839), 1081.
128 'Of 129 rheumatic cases treated by Dr Jules Cloquet, about 85 yielded to acupuncture. Of 34 published by others, 28 were cured. The writer of this article employed it in St. Thomas's Hospital, and published his results ... Of 42 cases, taken in succession as they stood in the hospital-books, 30 were found to have been cured: and the remaining 12 had clearly not been adapted for the remedy' (Elliotson, 'Acupuncture', 32).
129 Elliotson, 'Acupuncture', 32.
130 Churchill, *Treatise*, 22–3.
131 I have found no evidence of experimental investigations of acupuncture in Britain in the period from 1802 to 1830, although as has been noted above, clinical evaluations were performed and quantified by its British supporters. For more information on the transmission of medical and anatomical knowledge fron France to Britain during this period see Russell C. Maulitz, *Morbid Appearances: the Anatomy of Pathology in the Early Nineteenth Century* (Cambridge, 1987), Part II.

4 Networks and Innovations

1 James Morss Churchill, *Cases Illustrative of the Immediate Effects of Acupuncturation, in Rheumatism, Lumbago, Sciatica, Anomalous Muscular Diseases, And in Dropsy of the Cellular Tissue; Selected from various sources, and intended as an Appendix to the Author's Treatise on the Subject* (London, 1828), 22–3.
2 Churchill, *Cases*, 17.
3 Churchill, *Cases*, 22–3.
4 Churchill, *Cases*, 27.
5 Churchill, *Cases*, 26.
6 Churchill, *Cases*, 67.
7 Few were as articulate about their disgust as William Temple (see Chapter 2), but their sentiments seemed much the same.

8   See Chapter 3, text and notes.
9   Churchill, *Cases*, 44–5.
10   Churchill, *Cases*, 46.
11   As the cases of electricity, phrenology and mesmerism demonstrate, professional ambivalence towards amateur experimentation was common. See below, and Iwan Morus, *Frankenstein's Children: Electricity, Exhibition and Experiment in Early Nineteenth-Century London* (Princeton, 1998); Roger Cooter, *The Cultural Meaning of Popular Science: Phrenology and the Organisation of Consent in Nineteenth-Century Britain* (Cambridge, 1984); Alison Winter, *Mesmerized: Powers of Mind in Victorian Britain* (Chicago, 1998).
12   Churchill, *Cases*, 75
13   'London Medical Society. March 18th, 1833. Mr Kingdon, President. Rheumatism – Elaterium. Acupuncture', *Lancet* (23 March 1833): 817–18, at p. 817.
14   'London Medical Society', *Lancet*, 817.
15   'London Medical Society', *Lancet*, 817–18.
16   See James Morss Churchill, *A Treatise on Acupuncturation; Being a Description of a Surgical Operation Originally Peculiar to the Japonese and Chinese, and by them denominated Zin-King, Now Introduced into European Practice, with Directions for its Performance and Cases Illustrating its Success* (London, 1822), 71–2.
17   Scott, who was only ever identified by his (extremely common) last name and the location of his practice, has unfortunately proven completely untraceable beyond Churchill's brief reference to him – a reference periodically repeated in the medical journals' discussions of acupuncture's history. Tatam Banks referred to Paris in Banks, 'Observations on Acupuncturation', *The Edinburgh Medical and Surgical Journal*, 35 (1831): 323–8. J. Bossy claims that the first hospital experiments were performed in 1825 by Jules Cloquet at Saint Louis, but others in Paris were practising and experimenting on the technique at least a decade earlier. Bossy, 'The History of Acupuncture in the West', in Teizo Ogawa (ed.), *History of Traditional Medicine: Proceedings of the 1st and 2nd International Symposia on the Comparative History of Medicine – East and West* (Tokyo, 1986), 363–400, at p. 376.
18   See Steven Shapin and Simon Schaffer, *Leviathan and the Air-Pump: Hobbes, Boyle, and the Experimental Life: Including a Translation of Thomas Hobbes, Dialogus Physicus de Natura Aeris, by Simon Schaffer* (Princeton, 1985) for more on the culture of the witness, the audience, and the performance in science.
19   See Mike Saks, *Professions and the Public Interest: Medical Power, Altruism and Alternative Medicine* (London, 1995) for a comparatively full version of this argument in Britain. Conversely, in *Celestial Lancets: a History and Rationale of Acupuncture and Moxa* (Cambridge, 1980), Lu Gwei-Djen and Joseph Needham concluded from the frequent appearances of the term 'acupuncture' that the technique as well was commonly practised.
20   For more on these men and others involved in the medical reform movement, see Adrian Desmond, *The Politics of Evolution: Morphology, Medicine, and Reform in Radical London* (Chicago, 1989).
21   Churchill, *Treatise*, 23–4.
22   William Craig, 'Art. VII. Acupuncture in a case of Cancer', *Edinburgh Medical Journal, Combining the Monthly Journal of Medicine and the Edinburgh Medical and Surgical Journal*, 14 (1869): 617–20, at p. 619.

23 See Roy Porter 'The Physical Examination', in W. F. Bynum and Porter (eds), *Medicine and the Five Senses* (Cambridge, 1993), 179–97; Charles Rosenberg, 'The Therapeutic Revolution: Medicine, Meaning, and Social Change in 19th Century America', in Judith Leavitt and Ronald L. Numbers (eds), *Sickness and Health in America: Readings in the History of Medicine and Public Health*, 2nd edition (Madison, 1985), 39–52; and also his *The Care of Strangers: the Rise of America's Hospital System* (New York, 1987); Dorothy Porter and Roy Porter, *Patient's Progress; Doctors and Doctoring in Eighteenth-Century England* (Stanford, 1989); Dorothy Porter and Roy Porter, *In Sickness and in Health: the British Experience, 1650–1850* (London, 1988); Anne Digby, *Making a Medical Living: Doctors and Patients in the English Market for Medicine, 1720–1911* (Cambridge, 1994). As has been previously discussed, several practitioners explicitly credited patient demand for their use of acupuncture, and Churchill, at least, described several cases in which he applied the needles at patients' requests and against his own diagnostic judgement.

24 John Tatam Banks, 'Observations on Acupuncturation', 324

25 Churchill, *Cases*, 34.

26 Churchill, *Cases*, 74–5. Many of Churchill's cases, including this one, were taken directly from the medical periodicals. However, some contain extra details reported to him directly by the practitioners.

27 John Tatam Banks, 'Observations on Acupuncturation', 324.

28 Banks, 'Observations on Acupuncturation', 325–6.

29 Banks, 'Observations on Acupuncturation', 325–6.

30 This and the preceding quote are from Churchill, *Cases*, 72–3.

31 Winter, *Mesmerized*, 109–62. Winter points out that magical language was regularly used to describe science and technology in this period.

32 See Winter, *Mesmerized*. It is also possible that acupuncture's Chinese origins here acted to protect the therapy from connotations of French radicalism. Certainly it involved little physical contact with the patient, and so avoided the sexual scandal which plagued mesmeric practice.

33 Churchill, *Cases*, 46.

34 Churchill, *Cases*, 53. Sankey, the surgeon in question, enjoyed the luxury to perform controlled experiments on his patients by virtue of their poverty; one was a fisherman, the other unemployed because of his incapacitating pain.

35 Felix Vicq D'Azyr (ed.), *Médecine. Contentant, L'Hygiène; La Pathologie, La Séméiotique & La Nosologie; La Thérapeutique Ou Matière Médicale; La Médecine Militaire; La Médecine Vétérinaire; La Médecine Légale; La Jurisprudence De La Médecine &De La Pharmacie; La Biographie Médicale. C'est-a-dire, les Vies des Médecins Célèbres, avec des Notices de Leurs Ouvrages*, in Diderot and d'Alembert (eds), *Encyclopédie Méthodique*, Tome 1 (Paris, 1792), 185. In France, this overlapping of medicine and surgery would have worked in acupuncture's favour, given the forced marriage of the two disciplines in the Republic. See Russell Maulitz, *Morbid Appearances: the Anatomy of Pathology in the Early Nineteenth Century* (Cambridge, 1987).

36 Scottish doctors present something of an exception to this rule, perhaps because the close ties between Edinburgh and Paris enabled the younger generation of Scottish physicians and surgeons to absorb the boundary-crossing ethos of post-Revolutionary French medicine. See J. S. Jacyna, *Philosophic*

Whigs: Medicine, Science and Citizenship in Edinburgh 1789–1848 (London, 1994), especially his chapter on Pathology. He argues that surgery in Edinburgh was taught as more or less all-encompassing (in the French manner) but that it remained profoundly empirical in the sense that it was not taught as limited by theory or jargon.

37 This and preceding quotations, Churchill, *Cases*, 37.

38 Saks makes a strong-programme argument for this type of response in *Professions and the Public Interest*.

39 At St Thomas's, acupuncture was in fact practised by a physician, John Elliotson. However, Elliotson was strongly influenced by Parisian ideas, including the post-Revolutionary *rapprochement* between surgery and medicine, and the subsequent overlap between medical and surgical spheres and techniques of practice. See Russell Maulitz, *Morbid Appearances*.

40 'Hospital Reports: St Thomas's Hospital. Case of Chronic Rheumatism, cured by Acupuncture', *Lancet* (18 August 1826): 636–7, at p. 637, emphasis in the original.

41 'Hospital Reports: St Thomas's Hospital', *Lancet*, 637.

42 'Hospital Reports: St Thomas's Hospital. Efficacy of Acupuncturation in Rheumatism', *Lancet* (20 December 1826): 429–30, at p. 430, emphasis in the original.

43 Of course, there is a certain irony in the likening of needling to phlebotomy, given the eighteenth-century role of this comparison in hindering the adoption of acupuncture!

44 T. W. Wansbrough, 'Case of Rheumatism Successfully Treated by Acupuncturation', *Lancet* (21 June 1828): 366–7. This comment was drawn from a editorial introduction to Wansbrough's article at p. 366. A further editorial remark reminded the reader of an article published elsewhere (in the *Medico-Chirurgical Transactions*) in which the records of St Thomas's Hospital showed that of 42 cases of acupuncture taken in succession from the case books, 30 were cured and the other 12 were judged by John Elliotson to have been inappropriate for needling.

45 They did condemn homoeopathy, mesmerism, and metallic tractors, and their users.

46 Roger French and Andrew Wear, *Medicine and an Age of Reform* (London, 1991); Chris Lawrence (ed.), *Medical Theory, Surgical Practice: Studies in the History of Surgery* (London, 1992); Digby, *Making a Medical Living*.

47 Remember that the early connection drawn between the stethoscope and the needles (in this case, as mere novelties, unworthy of further consideration) was made by a speaker at the Royal College of Medicine. For further discussion of the medical response to the stethoscope, see Stanley J. Reiser, *Medicine and the Reign of Technology* (London, 1978), especially Chapter 2.

48 Erasmus Darwin, *Zoonomia*, Vol. 1 (London, 1794), 1–2.

49 Churchill, *Cases*, 2.

50 Churchill, *Cases*, 2.

51 Churchill, *Cases*, 3

52 James Morss Churchill, 'On Acupuncturation', *The London Medical Repository*, 19 (1823): 372.

53 Churchill, 'On Acupuncturation', 372.

54 Churchill, 'On Acupuncturation', 372.

55 'Quarterly Periscope or, Spirit of the Public Journals... Acupuncture (Mr Churchill)', *The Medico-Chirurgical Review, and Journal of Medical Science (Quarterly)*, 4 (1824): 956–7

56 Churchill, *Cases*, 3. The forum within which such attacks took place is unspecified.

57 Churchill, *Cases*, 3–4.

58 Churchill, *Cases*, 15–6.

59 Churchill, *Cases*, 15–16. For more on British perceptions of the French, see Linda Colley, *Britons: Forging the Nation, 1707–1837* (New Haven, 1992).

60 Churchill, *Cases*, 5.

61 Churchill, *Cases*, 5–6.

62 Churchill, *Cases*, 6–7.

63 Churchill, *Cases*, 7

64 In the ailments for which he recommended acupuncture, insertion points were commonly located near the site of pain, which may explain his success and the failures of his less discriminating emulators. There is no compelling evidence to suggest that Churchill was using Chinese acupuncture points deliberately or knowledgeably. However, using these sensations as the gauge of correct needle-placement would have inadvertently ensured a closer fit with Chinese body-maps.

65 Churchill, *Cases*, 16.

66 Churchill, *Cases*, 17.

67 Churchill, *Cases*, 20.

68 See Chapter 3 for a brief discussion of the origins of the second meaning of 'acupuncture' and for more on the singular needle.

69 John Renton, 'Observations on Acupuncture', *Edinburgh Medical and Surgical Journal*, 34 (1830): 100–1, at p. 100.

70 Renton, 'Observations on Acupuncture', 100.

71 Renton, 'Observations on Acupuncture', 101.

72 *Report of the Speeches Delivered at the Public Dinner Given at Pennycuik to John Renton, Esq. Surgeon. May 8 1835* (Edinburgh, 1835). At this dinner, published in celebration of the Parliamentary Reform Bill of 1832, Renton noted the involvement of the medical community in the politics of the day: 'I have been told that a medical man has nothing to do with politics. This was invariably the opinion of those whose political opinions were at variance with my own... Their sincerity, however, might well be questioned, considering what tools they made of medical men, whose names it would be invidious in me to mention.', (p. 5). He concluded: 'Remember, Gentlemen, we live in no ordinary times. The reform bill... was an epoch in a nation's history', (p. 6).

73 See Desmond, *Politics of Evolution*, 374–87.

74 Renton, 'Observations on Acupuncture', 101.

75 Renton, 'Observations on Acupuncture', 101.

76 See, for example, J. Bossy, 'History'; G. Lu and J. Needham, *Celestial Lancets*; G. Feucht, *Die Geschicte der Akupunktur in Europa* (Heidelberg, 1977); Roger Baptiste, *L'Acupuncture et son Histoire: Avantages et Inconvénients d'une Thérapeutique Millénaire* (Paris, 1962); and Daniel Geoffroy, *L'Acupuncture an France au XIXe siècle* (Paris, 1986) on the history of acupuncture in Europe, and particularly France.

77 John Elliotson, 'St Thomas Hospital. Clinical Lecture delivered by John Elliotson, MD, FRS, Physician to the Hospital, and Professor of the Principles and Practice of Medicine in the University of London. October 22nd, 1832', *Lancet* (3 November 1832): 161–7, at p. 167.

78 Elliotson, 'Clinical Lecture', 167, my emphasis.

79 Elliotson, 'Clinical Lecture', 167, my emphasis.

80 A Surgeon, *Animal Magnetism Delineated by its Professors. A Review of its History in Germany, France, and England. Reprinted from Number XIV of the British and Foreign Medical Review* (London, 1839), 3–4.

81 Again, see Desmond, *Politics of Evolution*, 382–97.

82 Dr Daser, 'Surgical Pathology and Therapeutics. "Reduction of a Strangulated Hernia, apparently effected by Acupuncture." ' *Edinburgh Medical and Surgical Journal*, 56 (1842): 549.

83 This class of periodicals ranged from the scholarly and elite *Journal of the Royal Asiatic Society* and its more accessible offshoot, the *Proceedings of the North China Branch of the Royal Asiatic Society*, to the professional *China Medical Journal*, and the missionary publication the *Anglo-Chinese Gleaner*.

84 See also Chapter 3.

85 James Henderson, 'Article V. The Medicine and Medical Practice of the Chinese', *Journal of the North-China Branch of the Royal Asiatic Society*, New Series, 7 (1864): 21–69, at p. 57. It is worth noting that the use of acupuncture in China during this period was by no means confined to inflammatory conditions – indeed, Henderson's description sounds more like a botched western acupuncture operation. It is possible that Henderson never witnessed a Chinese acupuncturist at work, or did not recognize the operation as acupuncture.

86 For a quick sketch of missionary and other medical activities in the Yunnan region of China, see Elisabeth Hsu, 'The Reception of Western Medicine in China: Examples from Yunnan', in Patrick Petitjean, Catherine Jami and Anne Marie Moulin (eds), *Science and Empires: Historical Studies about Scientific Development and European Expansion* (London, 1992), 89–101. For a more detailed analysis, see Bridie Andrews, 'The Making of Modern Chinese Medicine, 1895–1937' (PhD Dissertation, University of Cambridge, 1996) and Bridie Andrews, 'Tuberculosis and the Assimilation of Germ Theory in China, 1895–1937', *Journal of the History of Medicine and the Allied Sciences*, 52 (1997): 114–57.

87 Robert Hooper, *Lexicum Medicum; or Medical Dictionary; Containing an Explanation of the Terms in Anatomy, Physiology, Practice of Physic, Materia Medica, Chemistry, Pharmacy, Surgery, Midwifery, and the Various Branches of Natural Philosophy Connected with Medicine*, 4th edition (London, 1820), 16. This definition repeats that of the 1811 edition.

88 Robert Hooper, *Lexicon Medicum*, 7th edition revised by Grant Klein (London, 1839), 33.

89 Hooper, *Lexicon Medicum*, 7th edition, 33.

90 Hooper, *Lexicon Medicum*, 7th edition, 33.

91 Hooper, *Lexicon Medicum*, 7th edition, 33.

92 Hooper, *Lexicon Medicum*, 7th edition, 33.

93 John Elliotson, *The Principles and Practice of Medicine; Founded on the Most Extensive Experience in Public Hospitals and Private Practice; and as Developed in a*

Course of Lectures, Delivered at University College, London (London, 1839), 1018–25.

94   Shirley Palmer, *A Pentaglot Dictionary of the Terms Employed in Anatomy, Physiology, Pathology, Practical Medicine, Surgery, Obstetrics, Medical Jurisprudence, Materia Medica, Pharmacy, Medical Zoology, Botany, and Chemistry in Two Parts* (London, 1845), 12.

95   Robley Dunglison translated Baron Larrey's treatise on moxibustion, adding a long introduction describing the history of the technique and its 'sensible' use. In this introduction, he also mentioned acupuncture. Dunglison later emigrated to Philadelphia, where he became prominent in US medicine. See D. J. Larrey, *On the Use of Moxa as a Therapeutic Agent; Translated from French, with Notes and an Introduction Containing a History of the Substance, by Robley Dunglison* (London, 1822).

96   Robley Dunglison, *Medical Lexicon. A New Dictionary of Medical Science, Containing a Concise Account of the Various Subjects and Terms; with the French and Other Synonyms and Formulae for Various Officinal and Empirical Preparations,* 3rd edition (Philadelphia, 1842), 21.

97   The idea that acupuncture was used in some way in difficult pregnancies had a long European pedigree. Ten Rhyne reported that the Japanese needled the foetus to quiet it in cases where its movements endangered the mother, but he did not connect this with abortion – his tone implied that this action preserved both mother and infant. The mere idea of interfering with the uterus in this way seems to have horrified western doctors, and it was frequently cited as an example of Asian incompetence and anatomical ignorance. However, no other authors explicitly described it as a form of abortion.

98   Robley Dunglison, *New Remedies: Pharmaceutically and Therapeutically Considered,* 4th edition (Philadelphia, 1843), 45–53, at p. 45.

99   Dunglison, *New Remedies,* 46.

100   Dunglison, *New Remedies,* 47.

101   Dunglison, *New Remedies,* 49–50.

102   Robley Dunglison, *History of Medicine from the Earliest Ages to the Commencement of the Nineteenth Century. Arranged and Edited by Richard J. Dunglison* (Philadelphia, 1872). See in particular Chapter 7, 'Medicine of the Chinese and Japanese'. It begins with a section entitled, 'Causes of their imperfect civilization', and becomes steadily more negative throughout the chapter. Dunglison concluded that all of China's accurate knowledge and useful expertise was developed before the thirteenth century, and had its origins elsewhere: 'it is highly probable that they had previously had communication with the advanced nations of Europe, and that they had acquired from them some of their knowledge ... The notions which the Chinese possess regarding the structure of the body mainly rest on old [Greek] traditions; superstition preventing them from dissecting (pp. 72–3).

103   Robert Kemp Philip, *The Dictionary of Medical and Surgical Knowledge and Complete Practical Guide in Health and Disease for Families, Immigrants and Colonists,* Vol. 1 (London, 1864), 19.

104   Philip, *Dictionary,* 19.

105   Philip, *Dictionary,* 19.

106   S. O. Beeton, *Beeton's Medical Dictionary. A Safe Guide for Every Family. Defining in the Plainest Language, the Symptoms and Treatment of All Ailments,*

Illnesses and Diseases...And a Full Explanation of Medical and Surgical Terms (London, 1871), 4. Indeed, proponents of hypodermic injections and vaccination used acupuncture as evidence for the safety of subcutaneous needling (see, for example, A. N. Hill, 'Acupuncture the Best Method of Vaccination', *Annual Journal of Public Health*, 7 [1917]: 301).

107 Edwin Lankester (ed.), *Haydn's Dictionary of Popular Medicine and Hygiene; Comprising All Possible Self-Aids in Accident and Disease* (London, 1874), 8.

108 Lankester, *Haydn's Dictionary*, 8.

109 Robert Druitt, *The Surgeon's Vade Mecum, A Handbook of the Principles and Practice of Surgery* (London, 1839), 389.

110 William Fergusson, *A System of Practical Surgery* (London, 1842), 535. In his 'Chapter VI Operations on the Scrotum, Testicle, Prepuce and Penis', Fergusson noted that 'Some years ago this method of treatment attracted a good deal of attention; but as far as I can see it has undeservedly passed out of notice again – perhaps in consequence of the over-sanguine statements of those who advocated the plan.'

111 Annandale, *Surgical Appliances and Minor Operative Surgery* (Edinburgh, 1866), 52. Annandale was a FRCS of Edinburgh, Lecturer on Surgery and Assistant Surgeon to Edinburgh Royal Infirmary, and had been the Demonstrator of Anatomy at the University of Edinburgh.

112 John Eric Erichsen, *The Science and Art of Surgery. A Treatise on Surgical Injuries, Diseases and Operations*, 8th edition, Vol. 2 (London, 1884), 119.

113 Marcus Beck, 'Acupuncture', in Richard Quain (ed.), *A Dictionary of Medicine, Including General Pathology, General Therapeutics, Hygiene, and the Diseases Peculiar to Women and Children* (London, 1882), 12.

114 Beck, 'Acupuncture', 12.

115 Beck, 'Acupuncture', 12.

116 Beck, 'Acupuncture', 12.

117 Beck, 'Acupuncture': 12. For more on Baunsheidtism, see John Haller, 'Acupuncture in Nineteenth Century Western Medicine', *New York State Journal of Medicine*, 73 (1973): 1213–21. Haller concentrates on the role played by Baunscheidtism in American responses to acupuncture, advancing little direct evidence, but several suggestive coincidences. I have not found direct evidence of allusions of quackery in connection with acupuncture in Britain, perhaps because British medicine remained preoccupied with its French counterpart in this period. For more on the relationship between acupuncture and Baunscheidtism in Germany, see Bossy, 'History', 388–90.

118 See Haller, 'Acupuncture', 1213–21 for more detail; also C. F. Lallemand, *A Practical Treatise on the Causes, Symptoms, and Treatment of Spermatorrhea* (London, 1847); A Court Physician, *Reproductive Disorders, Spermatorrhagia, Exhausted Brain, Etc.* (London, 1876).

119 See, for example, Arnold and Sons, *Catalogue of Surgical Instruments* (London, 1885), 51. Acupuncture needles cost 1s. 6d. each, while ordinary suturing needles were 1s. 6d. for six.

120 Chauncy Puzey, 'Acupuncture', in Christopher Heath (ed.), *A Dictionary of Practical Surgery* (London, 1886), 24–6, at p. 24. Unfortunately, no biographical information about Puzey was included other than that he was a 'Hospital Surgeon'.

121 Puzey, 'Acupuncture', 24–5.
122 Puzey, 'Acupuncture', 24–5.
123 Puzey, 'Acupuncture', 24–5.
124 John Tatam Banks, 'The Treatment of Some Forms of Rheumatism and Neuralgic Affections', *Lancet* (14 June 1856): 652–3.
125 Banks, 'Treatment', 652–3.
126 Banks, 'Treatment', 652.
127 Banks, 'Treatment', 652, emphasis in the original.
128 Banks, 'Treatment', 652.
129 Banks, 'Treatment', 652–3.
130 Banks, 'Treatment', 653.
131 Banks, 'Treatment', 653, emphasis in the original.
132 Banks, 'Treatment', 653.
133 William Craig, 'Art. VII. Acupuncture in a Case of Cancer', *Edinburgh Medical Journal*, 14 (1869): 617–20, at p. 618. The *Edinburgh Medical Journal* was the fruit of a merger between the *Monthly Journal of Medicine* and the *Edinburgh Medical and Surgical Journal*.
134 This and all immediately preceding quotes are from Craig, 'Acupuncture', 618. This is his own summary of the argument presented in that text, and is an accurate reflection.
135 Craig, 'Acupuncture', 618.
136 Craig, 'Acupuncture', 619.
137 Craig, 'Acupuncture', 619.
138 T. Pridgin Teale, 'Clinical Essays, No. III. On the Relief of Pain and Muscular Disability by Acupuncture', *Lancet* (29 April 1871): 567–8, at p. 567.
139 Teale, 'On the Relief of Pain', 567.
140 Teale, 'On the Relief of Pain', 567.
141 Teale, 'On the Relief of Pain', 567.
142 Teale, 'On the Relief of Pain', 568.
143 Entries for October 1825, May 1826, November 1832 and May 1833 from 'Leeds General Infirmary Meetings of the Faculty 1824–1885', LGI Archives, MSS 5/1. Sadly, little material dating from this period has survived in the Infirmary's archives.
144 J. Brindley James, 'Treatment of Lumbago and Rheumatic Pains by the Percusso-Punctator. Read in the Section of Surgery at the Annual Meeting of the British Medical Association in Cardiff, 1885' (London, 1886), 1–16, at p. 3. The pamphlet sold for a shilling, and second edition was issued in 1897. His instrument was available through Arnold and Sons and other distributors, selling for 15s. in 1895. See Arnold and Sons, *Catalogue of Surgical Instruments and Appliances* (London, 1895), 757.
145 James, 'Treatment of Lumbago', 6.
146 James, 'Treatment of Lumbago', 6–7.
147 James, 'Treatment of Lumbago', 9–10.
148 James, 'Treatment of Lumbago', 13–14.
149 James, 'Treatment of Lumbago', 14–15.
150 James, 'Treatment of Lumbago', 5.
151 G. Lorimer, 'Acupuncture and its Application in the Treatment of Certain Forms of Chronic Rheumatism', *British Medical Journal* (21 November 1885): 956–8, at p. 956.

152 Lorimer, 'Acupuncture', 956–7.
153 Lorimer, 'Acupuncture', 957.
154 Lorimer, 'Acupuncture', 957.
155 Lorimer, 'Acupuncture', 957.
156 E. Valentine Gibson, 'An Analysis of One Thousand Cases of Primary Sci-
     atica, With Special Reference to the Treatment of One Hundred Cases by
     Acupuncture', *Lancet* (15 April 1893): 860–1.
157 Valentine Gibson, 'An Analysis', 860.
158 Valentine Gibson, 'An Analysis', 861.
159 Valentine Gibson, 'An Analysis', 861.
160 Valentine Gibson, 'An Analysis', 861.
161 Valentine Gibson, 'An Analysis', 861.
162 By 1909, a doctor at the Leeds General Infirmary was treating both rheuma-
     tism and osteoarthritis with salycilic acid, while the use of asepsis and anaes-
     thesia had allowed the return of paracentesis. Records 25, 44, 52, 53, 60, 84
     from 'Register of Case, Dr Griffith Ward. Vol. 1 1908–9', LGI Archives, MSS 78.
163 J. Brindley James, *Sciatica and its Treatment. Being a Paper Read in the Section of
     Medicine at the Annual Meeting of the British Medical Association, Sheffield, 1908*
     (London, 1908), 1.
164 James, *Sciatica*, 3.
165 Mike Saks, 'The Flight from Science? The Reporting of Acupuncture in
     Mainstream British Medical Journals from 1800–1990', *Complementary Med-
     ical Research*, 5 (1991): 178–82, Table 1 at p. 178.

Conclusions

1 C. A. Gordon, *An Epitome of the Reports of the Medical Officers to the Chinese
   Imperial Maritime Customs Service, from 1871 to 1882. With Chapters on the
   History of Medicine in China; Materia Medica; Epidemics; Famine; Ethnology; and
   Chronology in Relation to Medicine and Public Health* (London, 1884), 268–9.
2 See R. C. Crozier, *Traditional Medicine in Modern China: Nationalism and the
   Tensions of Cultural Change* (Cambridge, MA, 1968).
3 For more on the importance of correspondence in Chinese medicine, see
   Paul Unschuld, *Medicine in China: a History of Pharmaceutics* (Berkeley, 1986).
4 See Roger French and Andrew Wear (eds), *The Medical Revolution of the
   Seventeenth Century* (Cambridge, 1989).
5 See for example, Robeley Dunglison's last description of Chinese medicine
   in his *History of Medicine from the Earliest Ages to the Commencement of the
   Nineteenth Century* (Philadelphia, 1872). About the Chinese, he comments:
   '[T]heir anatomical information has been so very incorrect and confused as
   to scarcely deserve mention. A single glance at the plates given by Cleyer in
   his *Specimen Medicinae Sinicae*, will at once show their slight knowledge of
   the human organization . . . Their physiology is not less contemptible' (pp.
   73–4); and 'The other principles of Chinese medicine are equally devoid of
   rationality with their theory of the pulse' (p. 76); or John Wilson (then the
   Inspector of Naval Hospitals and Fleets), *Medical Notes on China* (London,
   1846): 'The healing art among the Chinese, with much pretension to learning and practical power, is in a very rude and inefficient state: it is, in fact, a

chaos of unfounded conceits, contradictory notions, and pompous phrases. Doctrinally, it has close analogy with the system of Pythagoras, as amplified, illustrated, and applied to medicine by Hippocrates; although it does not possess the coherence and methodical beauty which the former gave to his speculations, nor the keen observations of natural actions, close study of their relations, and acute practical precepts of the latter' (p. 233).

6   Lu Gwei-Djen and Joseph Needham, *Celestial Lancets: a History and Rationale of Acupuncture and Moxa* (Cambridge, 1980); Dorothy Rosenberg, 'Acupuncture and US Medicine: a Socio-Historical Study of the Response to the Availability of Knowledge' (PhD Dissertation, University of Pittsburgh, 1977).

7   John Haller, 'Acupuncture in Nineteenth Century Western Medicine', *New York State Journal of Medicine*, 73 (1973): 1213–21.

8   For a contemporary account, see Chauncy Puzey, 'Acupuncture', in Christopher Heath (ed.), *A Dictionary of Practical Surgery* (London, 1886), 24–6.

9   S. Hamilton et al., 'Anaesthesia by Acupuncture', *British Medical Journal* (April 1972): 352. See also Diamond, 'Acupuncture Anaesthesia: Western Medicine and Chinese Tradition', *Journal of the American Medical Association*, 71 (1971): 1560–3.

10   For innovation in the history of technology in general, see Weibe Bijker, Tom Hughes and Trevor Pinch (eds), *The Social Construction of Technological Systems: New Directions in the Sociology and History of Technology* (Cambridge, MA, 1987); for innovation in the hospital and in medical technology see Stuart Blume, *Insight and Industry: the Dynamics of Technological Change in Medicine* (Cambridge, MA, 1987); James Coleman et al., *Medical Innovation: a Diffusion Study* (Indianapolis, 1966); A. L. Greer, 'Medical Conservatism and Technological Acquisitiveness: the Paradox of Hospital Technology Adoptions', in J. A. Roth and S. B. Ruzek (eds), *The Social Impact of Medical Technology* (Greenwich, 1986), 185–235; Joel Howell, *Technology in the Hospital: Transforming Patient Care in the Twentieth Century* (Baltimore, 1995); Stanley J. Reiser and Michael Anbar (eds), *Medicine at the Bedside: Strategies for Using Technology in Patient Care* (Cambridge, 1984); Charles Rosenberg, *The Care of Strangers: the Rise of America's Hospital System* (New York, 1987). For innovation and medical specialism, see William Bynum, Chris Lawrence and Vivian Nutton (eds), *The Emergence of Modern Cardiology* (London, 1985). For a broader perspective on innovation in medicine, see Morris Vogel and Charles Rosenberg (eds), *The Therapeutic Revolution* (New York, 1979) and John Pickstone (ed.), *Medical Innovations in Historical Perspective* (Basingstoke, 1992).

11   Mary Fennell and Richard Warnecke, *The Diffusion of Medical Innovations: an Applied Network Analysis* (London, 1989), 2.

12   Even in the case of changing the patient-day by altering waking times, where initial investigations by Community Health Councils took place in response to persistent and nation-wide patient complaints about the organization of their days, Stocking did not regard patients as 'initiators'.

13   Bonnie Blair O'Connor's work on vernacular medicine in the US response to HIV/AIDS powerfully demonstrates the impact of consumer networks on medical provision and biomedical research and innovation. In such cases, as in the case of acupuncture, consumers can act as 'boundary spanners'.

14   The phrases 'alternative medicine' and 'complementary medicine' are hotly contested and freighted with heavy political and social connotations. For the

purposes of this discussion, it is worthwhile to note that those who support the use of unconventional medicine within the NHS typically refer to the desired practices as 'complementary', while those opposing such integration either from within or without the orthodox establishment prefer the term 'alternative'. When practised under the auspices of the NHS, it has proven virtually impossible for any heterodox therapy to be other than 'complementary' to biomedicine. See below, and Mike Saks, *Professions and the Public Interest: Medical Power, Altruism and Alternative Medicine* (London, 1995).

15 A 1995 WHICH report put British consumer-spending on alternative medicine at £60 million per year. See Consumer's Association, 'Health: Complementary Medicine. "Health Choice"', *WHICH* (November, 1995). Data on GPs and Health Authorities are drawn from, respectively, British Medical Association, *Complementary Medicine: New Approaches to Good Practice* (Oxford, 1993); and G. Cameron-Blackie, 'Complementary Therapies in the NHS', Research Paper 10, *National Association of Health Authorities and Trusts* (London, 1993).

16 Lewisham Hospital NHS Trust, 'Acupuncture Evaluation Project', unpublished draft document, 1993.

17 Janet Richardson, Anne-Marie Brennan and Robin Stott, 'Complementary Therapy Unit Proposal – September 1993' (Lewisham, 1993), 1.

18 Richardson et al., 'Proposal', 3. In Appendix 9, the authors list the BMA recommendations for the use of complementary therapies. These recommend that the GP 'delegate' the care of patients to complementary therapists in the same way that they might delegate 'routine tasks' to a practice nurse. Such recommendations clearly illustrate the BMA's vision of complementary medicine as a paramedical support profession.

19 Richardson et al., 'Proposal', Appendix 7.

20 Janet Richardson and Ann-Marie Brennan, 'Complementary Therapy in the NHS: Service Development in a Local District Hospital', *Complementary Therapies in Nursing and Midwifery*, 1(1995): 89–92, at p. 89.

21 Richardson and Brennan, 'Complementary Therapy', 92.

22 Janet Richardson, 'Complementary Therapies on the NHS: the Experience of a New Service', *Complementary Therapies in Medicine*, 3 (1995): 153–7.

23 Janet Richardson, *Complementary Therapy in the NHS: a Service Evaluation of the First Year of An Outpatient Service in a Local District General Hospital* (Lewisham, 1995), 7

24 Richardson, *Service Evaluation*, 15.

25 See Charles Webster, *The Great Instauration: Science, Medicine and Reform, 1626–1660* (London, 1975); Roy Porter and Dorothy Porter, *In Sickness and in Health: the British Experience, 1650–1850* (London, 1988); Dorothy Porter, and Roy Porter, *Patient's Progress: Doctors and Doctoring in Eighteenth-Century England* (Stanford, 1989).

26 See Chapter 2.

27 See Chapters 3 and 4.

28 Chinese philosophy and literature abound in cheap translations, and aspects of each have been incorporated into British popular culture. Flyers advertising acupuncture are decorated with yin/yang symbols and Chinese dragons, while Tai Chi classes are widely available in council leisure centres

and church halls. How thoroughly these cultural productions have permeated British life is, of course, more difficult to assess, but their presence is everywhere visible. The fact that only 25 per cent of patients attending the CTC in Lewisham first heard of acupuncture from their physician is, however, suggestive of a widespread awareness of the technique. Richardson, 'Service Evaluation', Appendix D.

29 'Complementary Therapy Service Operational Policy' (Lewisham, 1994), Appendix 4.

30 See for example, Felix Mann, *Acupuncture: the Ancient Chinese Art of Healing* (London, 1962) and G. Soulié de Morant, *L'Acupuncture Chinoise*, Vols 1 and 2 (Paris, 1939–41).

31 See Crozier, *Traditional Medicine*; John Z. Bowers and M. Purcell (eds), *Medicine and Society in China: Report of a Conference Sponsored Jointly by the National Library of Medicine and the Josiah Macy Jr. Foundation* (New York, 1974); Wong Chimin and Wu Lien-te, *History of Chinese Medicine* (Tientsin, 1932).

Bibliography

A Court Physician. *Reproductive Disorders, Spermatorrhagia, Exhausted Brain, Etc.* (London, 1876).

A Surgeon. *Animal Magnetism Delineated by its Professors. A Review of its History in Germany, France, and England. Reprinted from Number XIV of the British and Foreign Medical Review* (London, 1839).

Abel, Clarke. *Narrative of a Journey in the Interior of China, and of A Voyage to and from that Country, in the Years 1816 and 1817; Containing an Account of the Most Interesting Transactions of Lord Amherst's Embassy to the Court of Pekin, and Observations on the Countries which it Visited* (London, 1818).

'Acupuncturation', *Lancet* (9 November 1823): 200–1.

Adas, Michael. *Machines as the Measure of Men: Science, Technology, and Ideologies of Western Dominance* (Ithaca, 1989).

'Address', *Medical and Physical Journal*, 33 (1815): 1.

Adshead, S. A. M. *Material Culture in Europe and China, 1400–1800: the Rise of Consumerism* (Basingstoke, 1997).

'An Account of a Book, viz. Wilhelmi ten Ryne M.D. & c.... 1. Dissertatio de Arthritide. 2. Mantissa Schematica: 3. de Acupunctura.... Londini in 8vo 1683', *Philosophical Transactions*, 13 (1683): 222–35.

'An Historical Essay on the State of the Medical Sciences During the Last Six Months', *London Medical and Physical Journal*, 50 (1823): 1–66.

'Analytical Review, IV: *A Treatise on Acupuncturation: Being a Description of a Surgical Operation, originally peculiar to the Japonese [sic] and Chinese, now introduced into European Practice; with Directions for its Performance, and Cases illustrating its Success.* By James Morss Churchill, M.R.C.S., London, 1821', *London Medical Repository*, 17 (1822): 236–7.

Anderson, Aeneas. *A Narrative of the British Embassy to China, in the Years 1792, 1793, and 1794; with Accounts of the customs and manners of the Chinese; and a description of the country, towns, cities, &c.* 2nd edition (Dublin, 1796).

Andrews, Bridie. 'The Making of Modern Chinese Medicine, 1895–1937' (PhD dissertation, University of Cambridge, 1996).

Andrews, Bridie. 'Tuberculosis and the Assimilation of Germ Theory in China, 1895–1937', *Journal of the History of Medicine and the Allied Sciences*, 52 (1997): 114–57.

Andrews, Bridie and Andrew Cunningham. *Western Medicine as Contested Knowledge* (Manchester, 1997).

Annandale, Thomas. *Surgical Appliances and Minor Operative Surgery* (Edinburgh, 1866).

Anning, S. T. *The General Infirmary at Leeds. Volume I. The First Hundred Years 1767–1869* (London, 1963).

Anning, S. T. *The General Infirmary at Leeds. Volume II. The Second Hundred Years 1869–1965* (London, 1966).

Arnold and Sons. *Catalogue of Surgical Instruments* (London, 1885).

Arnold, David. *Colonizing the Body: State Medicine and Epidemic Disease in Nineteenth Century India* (Berkeley, 1993).

Arnold, David (ed.). *Imperial Medicine and Indigenous Societies* (Manchester, 1988).
'Art. XIV. [Review of Recent Work on Acupuncture]', *Edinburgh Medical and Surgical Journal*, 27 (1827): 190–200.
'Art. XIV. [Review of Recent Work on Acupuncture, continued.]', *Edinburgh Medical and Surgical Journal*, 27 (1827): 334–49.
Balfour, Frederic Henry. *Leaves from my Chinese Scrapbook* (London, 1887).
Banks, John Tatam. 'Observations on Acupuncturation', *Edinburgh Medical and Surgical Journal*, 35 (1831): 323–8.
Banks, John Tatam. 'The Treatment of some forms of Rheumatism and Neuralgic Affections', *Lancet* (14 June 1856): 652–3.
Baptiste, Roger. *L'Acupuncture et son Histoire: Avantages et Inconvénients d'une Thérapeutique Millénaire* (Paris, 1962).
Barfoot, Michael. 'Brunonianism Under the Bed: an Alternative to University Medicine in Edinburgh in the 1780s', in William Bynum and Roy Porter (eds), *Brunonianism in Britain and Europe* (London, 1988), 22–45.
Barrow, John. *Travels In China, Containing Descriptions, Observations, and Comparisons, Made and Collected in the Course of a Short Residence at the Imperial Palace of Yuen-Min-Yuen, and on a Subsequent Journey Through the Country from Pekin to Canton. In which it is Attempted to Appreciate the Rank that this Extraordinary Empire May Be Considered to Hold in the Scale of Civilized Nations* (London, 1804).
Bartlett, T. H. *Singular Listlessness: a Short History of Chinese Books and British Scholars* (London, 1989).
Beck, Marcus. 'Acupuncture', in Richard Quain (ed.), *A Dictionary of Medicine, Including General Pathology, General Therapeutics, Hygiene, and the Diseases Peculiar to Women and Children* (London,1882).
Beeton, S. O. *Beeton's Medical Dictionary. A Safe Guide for Every Family. Defining in the Plainest Language, the Symptoms and Treatment of All Ailments, Illnesses and Diseases . . . And a Full explanation of Medical and Surgical Terms* (London, 1871).
Berlioz, L. V. J. *Mémoires sur les Maladies Chroniques, les Évacuations Sanguines et l'Acupuncture* (Paris, 1816).
Breton, Jean-Baptiste. *China: Its Costume, Arts, Manufacture, &c. Edited Principally from the Originals in the Cabinet of M. Bertin; with Observations Explanatory, Historical, and Literary, by M. Breton*, 2nd edition (London, 1812).
Bhabha, Homi. *The Location of Culture* (London, 1994).
Bijker, Weibe, Tom Hughes and Trevor Pinch (eds). *The Social Construction of Technological Systems: New Directions in the Sociology and History of Technology* (Cambridge, MA, 1987).
Blume, Stuart. *Insight and Industry: the Dynamics of Technological Change in Medicine* (Cambridge, MA, 1987).
Bossy, J. 'The History of Acupuncture in the West', in Teizo Ogawa (ed.), *History of Traditional Medicine: Proceedings of the 1st and 2nd International Symposia on the Comparative History of Medicine – East and West* (Tokyo, 1986), 363–400.
Bowers, John Z. *Western Medical Pioneers in Feudal Japan* (Baltimore, 1970).
Bowers, John Z. and Robert Carrubba. 'The Doctoral Thesis of Engelbert Kaempfer: "On Tropical Diseases, Oriental Medicine and Exotic Natural Phenomenon"', *Journal of the History of Medicine and Allied Sciences*, 25 (1970): 270.
Bowers, John Z. and Elizabeth F. Purcell (eds). *Medicine and Society in China: Report of a Conference Sponsored Jointly by the National Library of Medicine and the Josiah Macy Jr. Foundation* (New York, 1974).

Bray, Francesca. 'A Deathly Disorder: Understanding Women's Health in Late Imperial China', in Don Bates (ed.), *Knowledge and the Scholarly Medical Traditions* (Cambridge, 1995), 235–50.

British Medical Association. *Complementary Medicine: New Approaches to Good Practice* (Oxford, 1993).

Brockway, L. H. *Science and Colonial Expansion: the Role of the British Royal Botanic Gardens* (New York, 1979).

Brou, A. 'Les Jésuites Sinologues de Pékin et leurs Editeurs de Paris', *Revue d'Histoire des Missions*, 11 (1934): 551–66.

Brown, E. Richard. 'Exporting Medical Education: Professionalization, Modernization, and Imperialism', *Social Science & Medicine*, 13A (1979): 585–95.

Bynum, William. 'Health, Disease and Medical Care', in R. Porter and G. S. Rousseau (eds), *The Ferment of Knowledge* (Cambridge, 1980).

Bynum, William, Chris Lawrence and Vivian Nutton (eds). *The Emergence of Modern Cardiology* (London, 1985).

Bynum, William, Stephen Lock and Roy Porter (eds). *Medical Journals and Medical Knowledge: Historical Essays* (London, 1992).

Bynum, William and Roy Porter (eds). *Brunonianism in Britain and Europe* (London, 1988).

Bynum, William and Roy Porter (eds). *Medical Fringe and Medical Orthodoxy 1750–1850* (London, 1987).

Cameron-Blackie, G. 'Complementary Therapies in the NHS', Research Paper 10, *National Association of Health Authorities and Trusts* (London, 1993).

Campbell, Donald. *Arabian Medicine and its Influence on the Middle Ages* (London, 1926).

Carlson, Eric and Meribeth Simpson. 'Models of the Nervous System in Eighteenth Century Psychiatry', *Bulletin for the History of Medicine*, 43 (1969): 100–15.

Carraro, A. 'Acupuncture of the Heart in Apparent Death', *Provincial Medical and Surgical Journal*, 2 (1841): 140.

Carrubba, Robert and John Z. Bowers, 'The Western World's First Detailed Treatise on Acupuncture: Willem Ten Rhijne's *De Acupunctura*', *Journal of the History of Medicine*, 29 (1974): 371–98.

Carter, Harry William. 'A General Report of the Medical Diseases Treated in the Kent and Canterbury Hospital, from January 1st to July 1st, 1823, with a Particular Account of the More Important Cases', *London Medical Repository*, 19 (1823): 386–402.

Carter, Harry William. 'Case No. XVIII "On Acupunctura"', *London Medical Repository*, 19 (1823): 455–6.

Chaudhuri, K. N. *The Trading World of Asia and the English East India Company, 1660–1760* (Cambridge, 1978).

Cheyne, George. *An Essay on Regimen, Together with Five Discourses, Medical, Moral and Philosophical: Serving to Illustrate the Principles and Theory of Philosophical Medicin [sic] and Point out Some of its Moral Consequences* (London, 1740).

Churchill, James Morss. 'On Acupuncturation', *London Medical Repository*, 19 (1823): 372.

Churchill, James Morss. *A Treatise on Acupuncturation; Being a Description of a Surgical Operation Originally Peculiar to the Japonese and Chinese, and by Them Denominated Zin-King, Now Introduced into European Practice, with Directions for its Performance and Cases Illustrating its Success* (London, 1821).

Churchill, James Morss. *Cases Illustrative of the Immediate Effects of Acupunctura-tion, in Rheumatism, Lumbago, Sciatica, Anomalous Muscular Diseases, And in Dropsy of the Cellular Tissue; Selected from Various Sources, and Intended as an Appendix to the Author's Treatise on the Subject* (London, 1828).

Clarke, Edwin and L. Stephen Jacyna. *Nineteenth-Century Origins of Neuroscientific Concepts* (Berkeley, 1987).

Codell, Julie and Dianne Sachko Macleod. *Orientalism Transposed: the Impact of the Colonies on British Culture* (Aldershot, 1998).

Cohen, Huguette. 'Diderot and China', *Studies on Voltaire*, 242 (1986): 219–32.

Coleman, James et al. *Medical Innovation: a Diffusion Study* (Indianapolis, 1966).

Coley. 'A Case of Tympanites, in an Infant, Relieved by the Operation of the Paracentesis. With Remarks on the Case; and a Critical Analysis of the Senti-ments of the Principal Authors who have Written on the Disease. To which is Subjoined an Account of the Operation of the Acupuncture, as Practised by the Japanese in the Diseases Analogous to the Tympany', *Medical and Surgical Journal*, 7 (1802): 235–6.

Colley, Linda. *Britons: Forging the Nation, 1707–1837* (New Haven, 1992).

Consumer's Association. 'Health: Complementary Medicine. "Health Choice"', *WHICH* (November, 1995).

Cooter, Roger. *The Cultural Meaning of Popular Science: Phrenology and the Organiza-tion of Consent in Nineteenth-Century Britain* (Cambridge, 1984).

Cooter, Roger (ed.). *Studies in the History of Alternative Medicine* (London, 1988).

Craig, William. 'Art. VII. Acupuncture in a Case of Cancer', *Edinburgh Medical Journal, Combining the Monthly Journal of Medicine and the Edinburgh Medical and Surgical Journal*, 14 (1869): 617–20.

Cranmer-Byng, J. L. (ed.). *An Embassy to China: Being the Journal kept by Lord Mac-artney During his Embassy to the Emperor Ch'ien-lung 1793–1794* (London, 1962).

Cranmer-Byng, J. L. and Trevor Levere. 'A Case Study in Cultural Collision: Scientific Apparatus in the Macartney Embassy to China, 1793', *Annals of Science*, 38 (1981): 503–25.

'Critical Analysis of English and Foreign Literature Relative to the Various Branches of Medical Science. Division II', *London Medical and Physical Journal*, 48 (1822): 518.

Crozier, R. C. *Traditional Medicine in Modern China: Nationalism and the Tensions of Cultural Change* (Cambridge, MA, 1968).

Dale, Ralph and Yan Cheng. 'An Outline History of Chinese Acupuncture: the Main Developments, Contributors and Publications', *American Journal of Acu-puncture*, 21 (1993): 355–93.

Darnton, Robert. *Mesmerism and the End of the Enlightenment in France* (Cambridge, MA, 1968).

Darwin, Erasmus. *Zoonomia; or the Laws of Organic Life*, 3rd edition (London, 1801).

Daser. 'Surgical Pathology and Therapeutics. "Reduction of a Strangulated Hernia, apparently effected by Acupuncture"', *Edinburgh Medical and Surgical Journal*, 56 (1842): 549.

Davis, John. *The Chinese: a General Description of the Empire of China and Its Inhabitants* (London, 1837)

de la Roche, F. and P. Petit-Radel (eds). 'Chirurgie', in Diderot and d'Alembert (eds), *Encyclopédie Méthodique* (Paris, 1792).

Desmond, Adrian. *The Politics of Evolution: Morphology, Medicine, and Reform in Radical London* (Chicago, 1992).

Diamond. 'Acupuncture Anaesthesia: Western Medicine and Chinese Tradition', *Journal of the American Medical Association*, 71 (1971): 1560–3.

Digby, Anne. *Making a Medical Living: Doctors and Patients in the English Market for Medicine, 1720–1911* (Cambridge, 1994).

Douglas, John. *A Short Dissertation on the Gout. Wherein the Universal Fear of Doing Anything to Ease or Cure It (Instilled in People's Heads by Both Ancient and Modern Writers) will be Proved to be a Mere Bug-Bear, a Groundless Supposition, a Vulgar Error, & c. and a Safe Method of Relieving the Most Violent Pains, Shortening the Fit, and Lengthening the Intervals, will be Proposed, and Confirmed by Several Cases* (London, 1741).

Druitt, Robert. *The Surgeon's Vade Mecum, A Handbook of the Principles and Practice of Surgery* (London, 1839).

DuHalde, J.-B. *A Description Of The Empire Of China And Chinese-Tartary, Together With The Kingdoms Of Korea And Tibet: Containing The Geography And History (Natural As Well As Civil) Of Those Countries. Enrich'd With General And Particular Maps, And Adorned With A Great Number Of Cuts. From The French Of P. J. B. DuHalde, Jesuit: With Notes Geographical, Historical, And Critical; And Other Improvements, Particularly In The Maps, By The Translator* (London, 1741).

DuHalde et al. (eds). *Lettres Édifiantes et Curieuses Écrits Des Missions Etrangères* (Paris, 1702–1776).

Dujardin, F. *Histoire de la Chirurgie, Depuis son Origine Jusqu'à nos Jours*, Tome 1 (Paris, 1774).

Dunglison, Robley. *History of Medicine from the Earliest Ages to the Commencement of the Nineteenth Century. Arranged and Edited by Richard J. Dunglison* (Philadelphia, 1872).

Dunglison, Robley. *Medical Lexicon. A New Dictionary of Medical Science, Containing a Concise Account of the Various Subjects and Terms; with the French and Other Synonyms and Formulae for Various Officinal and Empirical Preparations*, 3rd edition (Philadelphia, 1842).

Dunglison, Robley. *New Remedies: Pharmaceutically and Therapeutically Considered.* 4th edition (Philadelphia, 1843).

Elliotson, John. 'Acupuncture', in John Forbes, Alexander Tweedie and John Conolly (eds), *Cyclopaedia of Practical Medicine; Comprising Treatises on the Nature and Treatment of Diseases, Materia Medica and Therapeutics, Medical Jurisprudence, etc. etc.*, Vol. 1 (London, 1833), 32–3.

Elliotson, John. 'St Thomas Hospital. Clinical Lecture delivered by John Elliotson, MD, FRS, Physician to the Hospital, and Professor of the Principles and Practice of Medicine in the University of London. October 22nd, 1832', *Lancet* (3 November 1832): 161–7.

Elliotson, John. *The Principles and Practice of Medicine; Founded on the Most Extensive Experience in Public Hospitals and Private Practice; and as Developed in a Course of Lectures, Delivered at University College, London; With Notes and Illustrations by Nathaniel Rogers* (London, 1839).

Ellis, Henry. *Journal of the Proceedings of the Late Embassy to China; Comprising a Correct Narrative of the Public Transactions of the Embassy, of the Voyage to and from China, and of the Journey from the Mouth of the Pei-ho to the Return to Canton.*

Interspersed with Observations upon the Face of the Country, the Polity, Moral Character, and Manners of the Chinese Nation (London, 1817).

Engelstein, Laura. 'Morality and the Wooden Spoon: Russian Doctors View Syphilis, Social Class, and Sexual Behavior, 1895–1905', in C. Gallagher and T. Laqueur (eds), *The Making of the Modern Body: Sexuality and Society in the Nineteenth Century* (Berkeley, 1987), 169–208.

Erichsen, John Eric. *The Science and Art of Surgery. A Treatise on Surgical Injuries, Diseases and Operations*, 8th edition (London, 1884).

Farquhar, Judith, *Knowing Practice: the Clinical Encounter of Chinese Medicine* (Boulder, CO, 1994).

Fennell, Mary and Richard Warnecke. *The Diffusion of Medical Innovations: an Applied Network Analysis* (London, 1989).

Fergusson, William. *A System of Practical Surgery* (London, 1842).

Feucht, G. *Die Geschicte der Akupunktur in Europa* (Heidelberg, 1977).

Finch, Frederic. 'Case of Anascara in which Acupuncturation was Successfully Employed, and the Fluid Discharged by it', *London Medical Repository*, 19 (1823): 205–6.

Finch, Frederic. 'Case of Trismus, & c., approaching to Tetanus, Supervening to a Lacerated Wound, Successfully Treated by Acupuncturation', *London Medical Repository*, 19 (1823): 403–4.

Floyer, Sir John. *The Physician's Pulse-Watch; or, an Essay to Explain the Old Art of FEELING the PULSE, and to Improve it by the Help of a Pulse-Watch . . . To which is Added, An Extract out of Andrew Cleyer, Concerning the Chinese Art of Feeling the Pulse* (London, 1701).

Forbes, John, Alexander Tweedie and John Conolly (eds). *Cyclopaedia of Practical Medicine; Comprising Treatises on the Nature and Treatment of Diseases, Materia Medica and Therapeutics, Medical Jurisprudence, etc. etc.* (London, 1833).

'Foreign Department. Analysis of Foreign Medical Journals. Revue Médicale – June 1825. *On the Electro-magnetic phenomena Observed in Acupuncture*. By M. Pouillet', *Lancet* (6 August 1825): 152–3.

Forsyth, J. S. *The New London Medical and Surgical Dictionary* (London, 1826).

Fortune, Robert. *A Journey to the Tea Countries of China* (London, 1852).

Fothergill, Samuel. 'Address', *London Medical and Physical Journal*, 33 (1815): 1.

Foust, Clifford. *Rhubarb: the Wondrous Drug* (Princeton, 1992).

French, Roger and Andrew Wear (eds). *The Medical Revolution of the Seventeenth Century* (Cambridge, 1989).

Gallagher, C. and T. Laqueur (eds). *The Making of the Modern Body: Sexuality and Society in the Nineteenth Century* (Berkeley, 1987).

Garlick, T. *An Essay on the Gout . . . The Remedies, Both Internal and External Faithfully Publish'd in English, Without Reserve, for the Benefit of All Such as Now Do (or Hereafter May) Suffer by that Disease* (London, 1729).

Geoffroy, Daniel. *L'Acupuncture en France au XIXe siècle* (Paris, 1986).

Gibson, E. Valentine. 'An Analysis of One Thousand Cases of Primary Sciatica, with Special Reference to the Treatment of One Hundred Cases by Acupuncture', *Lancet* (15 April 1893): 860–1.

Gillan, Hugh. 'Mr Gillan's Observations on the State of Medicine, Surgery, and Chemistry in China', in J. L. Cranmer-Byng (ed.), *An Embassy to China: Being the Journal kept by Lord Macartney During his Embassy to the Emperor Ch'ien-lung 1793–1794* (London, 1962), 279–90.

Gooding, David, Simon Schaffer and Trevor Pinch (eds). *The Uses of Experiment: Studies in the Natural Sciences* (Cambridge, 1989).

Gordon, C. A. *An Epitome of the Reports of the Medical Officers to the Chinese Imperial Maritime Customs Service, from 1871 to 1882. With Chapters on the History of Medicine in China; Materia Medica; Epidemics; Famine; Ethnology; and Chronology in Relation to Medicine and Public Health* (London, 1884).

Granshaw, Lindsay. ' "Upon This Principle I Have Based A Practice": The Development and Reception of Antisepsis in Britain, 1890–1914', in John Pickstone (ed.), *Medical Innovations in Historical Perspective* (New York, 1992), 17–46.

Greenhalgh, Paul. *Ephemeral Vistas: the Expositions Universelles, Great Exhibitions and World's Fairs 1851–1939* (Manchester, 1988).

Greer, A. L. 'Medical Conservatism and Technological Acquisitiveness: the Paradox of Hospital Technology Adoptions', in J. A. Roth and S. B. Ruzek (eds), *The Social Impact of Medical Technology* (Greenwich, 1986), 185–235.

Guha, Ranajit. *Subaltern Studies: Writings on South Asian History and Society* (Delhi, 1981).

Guy, Basil. *The French Image of China, before and after Voltaire* (Geneva, 1963).

Haller, John. 'Acupuncture in Nineteenth Century Western Medicine', *New York State Journal of Medicine*, 73 (1973): 1213–21.

Hamilton, S. et al. 'Anaesthesia by Acupuncture', *British Medical Journal* (April 1972): 352.

Harrison, J. F. C. 'Early Victorian Radicals and the Medical Fringe', in William Bynum and Peters (eds), *Medical Fringe and Medical Orthodoxy, 1750–1850* (London, 1987), 198–215.

Heister, Lorenz. *A General System of Surgery in Three Parts* (London, 1743).

Henderson, James. 'Article V. The Medicine and Medical Practice of the Chinese', *Journal of the North-China Branch of the Royal Asiatic Society*, New Series, 7 (1864): 21–69.

Hevia, James. *Cherishing Men from Afar: Qing Guest Ritual and the Macartney Embassy of 1793* (London, 1995).

Hill, A. N. 'Acupuncture the Best Method of Vaccination', *Annual Journal of Public Health*, 7 (1917): 301.

Hooper, Robert. *Lexicon Medicum; or Medical Dictionary Containing an Explanation of the Terms in Anatomy, Physiology, Practice of Physic, Materia Medica, Chemistry, Pharmacy, Surgery, Midwifery, and the Various Branches of Natural Philosophy Connected with Medicine*, 7th edition revised by Grant Klein (London, 1839).

Hooper, Robert. *Lexicon Medicum; or Medical Dictionary Containing an Explanation of the Terms in Anatomy, Physiology, Practice of Physic, Materia Medica, Chemistry, Pharmacy, Surgery, Midwifery, and the Various Branches of Natural Philosophy Connected with Medicine*, 4th edition (London, 1820).

'Hospital Reports: St Thomas's Hospital. Case of Chronic Rheumatism, cured by Acupuncture', *Lancet* (18 August 1826): 636–7.

'Hospital Reports: St Thomas's Hospital. Efficacy of Acupuncturation in Rheumatism', *Lancet*, 1826–7, No. 174 (20 December 1826): 429–30.

Howell, Joel. *Technology in the Hospital: Transforming Patient Care in the Twentieth Century* (Baltimore, 1995).

Hsu, Elisabeth. 'Outline of the History of Acupuncture in Europe', *Journal of Chinese Medicine*, 29 (1989): 28–32.

Hsu, Elisabeth. 'The Reception of Western Medicine in China: Examples from Yunnan', in Patrick Petitjean, Catherine Jami and Anne Marie Moulin (eds), *Science and Empires: Historical Studies about Scientific Development and European Expansion* (London, 1992), 89–101.

Ingram, Dale. *An essay on the Cause and Seat of the Gout: in which the Opinions of Several Authors are Considered, and Some External Operations Recommended* (Reading, 1743).

Ingram, Richard. *The Gout... to which is Prefixed an Essay Pointing Out the Progressive Symptoms and Effects and the Reasons Why the Gout Was Not Heretofore Regularly Treated and Cured* (London, 1768).

Inkster, I. 'Marginal Men: Aspects of the Social Role of the Medical Community in Sheffield, 1790–1850', in J. H. Woodward and D. Richards (eds), *Health Care and Popular Medicine in Nineteenth Century England* (New York, 1977).

Inkster, I. and J. Morrell. *Metropolis and Province: Science in British Culture, 1780–1850* (Philadelphia, 1983).

Jacyna, L. Stephen. '"Mr Scott's Case": a View of London Medicine in 1825', in Roy Porter (ed.), *The Popularization of Medicine 1650–1850* (London, 1992), 252–86.

Jacyna, L. Stephen. *Philosophic Whigs: Medicine, Science and Citizenship in Edinburgh 1789–1848* (London, 1994).

James, J. Brindley. *Sciatica and its Treatment. Being a Paper Read in the Section of Medicine at the Annual Meeting of the British Medical Association, Sheffield, 1908* (London, 1908).

James, J. Brindley. *Treatment of Lumbago and Rheumatic Pains by the Percusso-Punctator. Read in the Section of Surgery at the Annual Meeting of the British Medical Association in Cardiff, 1885* (London, 1886).

James, R. *A Medicinal Dictionary; Including Physic, Surgery, Anatomy, Chymistry, and Botany, in all their Branches Relative to Medicine... and an Introductory Preface, Tracing the Progress of Physic, and Explaining the Theories which have Principally Prevail'd in All the Ages of the World* (London, 1743).

Jardine, Nick, James Secord and Emma Spary (eds). *Cultures of Natural History* (Cambridge, 1996).

Jewson, Nicholas. 'Medical Knowledge and the Patronage System in 18th Century England', *Sociology*, 8 (1974): 369–85.

Jewson, Nicholas. 'The Disappearance of the Sick-Man from Medical Cosmology, 1770–1870', *Sociology*, 10 (1976): 225–44.

Jordanova, L. *Sexual Visions: Images of Gender in Science and Medicine between the 18th and 20th centuries* (Madison, 1989).

Kaempfer, Engelbert. *The History of Japan: Giving an Account of the Antient and Present State and Government of that Empire; of Its Temples, Palaces, Castles, and other Buildings; of Its Metals, Minerals, Trees, Plants, Animals, Birds, and Fishes; of the Chronology and Succession of the Emperors, Ecclesiastical and Secular; of the Original Descent, Religions, Customs, and Manufactures of the Natives, and of their Commerce with the Dutch and Chinese. Together with a Description of the Kingdom of Siam* (London, 1728).

Khairallah, A. A. *Outline of Arabic Contributions to Medicine* (Beirut, 1946).

Kidd, Samuel. *China, or, Illustrations of the Symbols, Philosophy, Antiquities, Customs, Superstitions, Laws, Government, Education, and Literature of the Chinese* (London, 1841).

Koerner, Lisbet. 'Purposes of Linnaean Travel: a Preliminary Research Report', in David Miller and Hans Reill (eds), *Visions of Empire: Voyages, Botany, and Representations of Nature* (Cambridge, 1996), 117–52.

Kopf, David. *British Orientalism and the Bengal Renaissance: the Dynamics of Indian Modernization, 1773–1835* (Berkeley, 1969).

Kumar, A. *Medicine and the Raj: British Medical Policy in India, 1835–1911* (London, 1998).

Kuriyama, Shigehisa. 'Between Mind and Eye: Japanese Anatomy in the Eighteenth Century', in C. Leslie and A. Young (eds), *Paths to Asian Medical Knowledge* (Berkeley, 1992), 21–43.

Kuriyama, Shigehisa. 'Interpreting the History of Bloodletting', *Journal of the History of Medicine and Allied Sciences*, 50 (1995): 11–46.

Kuriyama, Shigehisa. 'Visual Knowledge in Classical Chinese Medicine', in Don Bates (ed.), *Knowledge and the Scholarly Medical Traditions* (Cambridge, 1995), 205–34.

Lallemand, C. F. *A Practical Treatise on the Causes, Symptoms, and Treatment of Spermatorrhea* (London, 1847).

Lankester, Edwin (ed.). *Haydn's Dictionary of Popular Medicine and Hygiene; Comprising All Possible Self-Aids in Accident and Disease* (London, 1874).

Lansbury, Coral. *The Old Brown Dog: Women, Workers, and Vivisection in Edwardian England* (Madison, 1985).

Laqueur, Thomas. *Making Sex: Body and Gender from the Greeks to Freud* (Cambridge, MA, 1990).

Larrey, D. J. *On the Use of Moxa as a Therapeutical Agent;Translated from French, with Notes and an Introduction Containing a History of the Substance, by Robley Dunglison* (London, 1822).

Lawrence, Chris. 'Cullen, Brown, and the Poverty of Essentialism', in William Bynum and Roy Porter (eds), *Brunonianism in Britain and Europe* (London, 1988), 1–21.

Lawrence, Chris. 'Democratic, Divine and Heroic: the History and Historiography of Surgery', in Chris Lawrence (ed.), *Medical Theory, Surgical Practice: Studies in the History of Surgery* (London, 1992), 1–47.

Lawrence, Chris. 'Incommunicable Knowledge: Science, Technology and the Clinical Art in Britain, 1850–1914', *Journal of Contemporary History*, 20 (1985): 503–20.

Lawrence, Chris. 'Ornate Physicians and Learned Artisans: Edinburgh Medical Men, 1726–1776', in William Bynum and Roy Porter (eds), *William Hunter and the Eighteenth-Century Medical World* (Cambridge, 1985), 153–76.

Lay, G. Tradescant. 'Minutes of the First Annual Meeting of the Medical Missionary Society in China', in *The First and Second Reports of the Medical Missionary Society in China: With Minutes of Proceedings, Hospital Reports, &c.* (Macao, 1841), 3–11.

Lay, G. Tradescant. *The Chinese as They Are: Their Moral, Social and Literary Characters; A New Analysis of the Language; with Succinct Views of Their Principal Arts and Sciences* (London, 1841).

Leclerc, Lucien. *Histoire de la Médecine Arabe: Exposé Complet des Traductions du Grec; les Science en Orient, leur Transmission à l'Occident par les Traductions Latines* (New York, 1961).

LeComte, Louis. *A Compleat History of the Empire of China: Being the Observations of Above Ten Years Travels through that Country... A New Translation from the Best Paris Edition* (London, 1739).

Lesch, J. E. *Science and Medicine in France: the Emergence of Experimental Physiology 1790–1855* (Cambridge, 1984).

Leslie, Charles and Allan Young (eds), *Paths to Asian Medical Knowledge* (Berkeley, 1992).

Linebaugh, Peter. 'The Tyburn Riots against the Surgeons', in D. Hay et al., *Albion's Fatal Tree* (London, 1975), 65–118.

'London Medical Society. March 18th, 1833. Mr Kingdon, President. Rheumatism. – Elaterium. Acupuncture', *Lancet* (23 March 1833): 817–18.

Lorimer, G. 'Acupuncture and its Application in the Treatment of Certain Forms of Chronic Rheumatism', *British Medical Journal* (21 November 1885): 956–8.

Loudon, Jean and Irvine Loudon. 'Medicine, Politics and the Medical Periodical, 1800–1850', in William Bynum, Stephen Lock and Roy Porter (eds), *Medical Journals and Medical Knowledge: Historical Essays* (London, 1992), 49–69.

Lowe, Lisa. *Critical Terrains: British and French Orientalisms* (Ithaca, 1991).

Lu Gwei-Djen and Joseph Needham. *Celestial Lancets: a History and Rationale of Acupuncture and Moxa* (Cambridge, 1980).

Ma, Kan-Wen. 'East–West Medical Exchange and Their Mutual Influence', in R. Hayhoe (ed.), *Knowledge Across Cultures, Universities East and West* (Canada, 1993): 154–81.

Mackay, David, 'Agents of Empire: the Banksian Collectors and Evaluation of New Lands', in Miller and Reill (eds), *Visions of Empire: Voyages, Botany and Representations of Nature* (Cambridge, 1996), 38–57.

MacKenzie, John (ed.). *Imperialism and Popular Culture* (Manchester, 1986).

MacLeod, Roy and Milton Lewis (eds). *Disease, Medicine and Empire: Perspectives on Western Medicine and the Experience of European Expansion* (London, 1988).

Mann, Felix. *Acupuncture: the Ancient Chinese Art of Healing* (London, 1962).

Marshall, P. J. 'Britain and China in the Late Eighteenth Century', in Robert A Bickers (ed.), *Ritual and Diplomacy: the Macartney Mission to China, 1792–1794* (London, 1993), 11–29.

Maulitz, Russell C. *Morbid Appearances: the Anatomy of Pathology in the Early Nineteenth Century* (Cambridge, 1987).

Maulitz, Russell C. 'The Pathological Tradition', in Roy Porter and William Bynum (eds), *Routledge Companion to the History of Medicine*, Vol. 1 (London, 1995), 169–91.

McKendrick, Neil, John Brewer and J. H. Plumb. *The Birth of a Consumer Society: the Commercialization of 18th century England* (London, 1982).

Medhurst, W. H. *The Foreigner in Far Cathay* (London, 1872).

'Medical and Philosophical Intelligence. "State of Medicine in China; by M. Page, of Orleans"', *London Medical and Physical Journal*, 33 (1815): 247–8.

'Medical and Physical Intelligence. Physiology. 2. On the Galvanic Phenomena which accompany the Acupuncturation', *London Medical and Physical Journal*, 53 (1825): 434.

Miller, David and Peter Reill (eds). *Visions of Empire: Voyages, Botany and Representations of Nature* (Cambridge, 1996).

Modern Part of the Universal History (London, 1759).

Morse, H. B. *The Chronicles of the East India Company, Trading to China, 1635–1834* (Oxford, 1926–9).

Morse, William. *Clio Medica XI: Chinese Medicine* (New York, 1934).

Morus, Iwan. *Frankenstein's Children: Electricity, Exhibition and Experiment in Early Nineteenth-Century London* (Princeton, 1998).

Mungello, D. E. *Curious Land: Jesuit Accommodation and the Origins of Sinology* (Honolulu, 1989).

Murray, Hugh et al. *An Historical and Descriptive Account of China; Its Ancient and Modern History, Language, Literature, Religion, Government, Industry, Manners, and Social State; Intercourse with Europe from the Earliest Ages; Missions and Embassies to the Imperial Court; British and Foreign Commerce; Directions to Navigators; State of Mathematics and Astronomy; Survey of its Geography, Geology, Botany, and Zoology* (Edinburgh, 1836).

Needham, Joseph. *The Grand Titration: Science and Society in East and West* (Toronto, 1969).

Nicholls, Phillip. 'Homoeopathy in Britain after the Mid-Nineteenth Century', in Mike Saks (ed.), *Alternative Medicine in Britain* (Oxford, 1992), 77–89.

Nicholson, Malcolm. 'Giovanni Battista Morgagni and Eighteenth-Century Physical Examination', in Chris Lawrence (ed.), *Medical Theory, Surgical Practice: Studies in the History of Surgery* (London, 1992), 101–34.

Nicholson, Malcolm. 'The Introduction of Percussion and Stethoscopy to Early Nineteenth-Century Edinburgh', in William Bynum and Roy Porter (eds), *Medicine and the Five Senses* (Cambridge, 1993), 134–53

Palmer, Shirley. *A Pentaglot Dictionary of the Terms Employed in Anatomy, Physiology, Pathology, Practical Medicine, Surgery, Obstetrics, Medical Jurisprudence, Materia Medica, Pharmacy, Medical Zoology, Botany, and Chemistry in 2 parts* (London, 1845).

'Pelletan on the Galvanic Phenomena of Acupuncturation', *Anderson's Quarterly Journal of the Medical Sciences*, 2 (1825): 428.

Peterson, M. Jeanne. *The Medical Profession in Mid-Victorian London* (Berkeley, 1978).

Petitjean, Patrick, Catherine Jami and Anne Marie Moulin (eds). *Science and Empires: Historical Studies about Scientific Development and European Expansion* (London, 1992).

Philip, Robert Kemp. *The Dictionary of Medical and Surgical Knowledge and Complete Practical Guide in Health and Disease for Families, Immigrants and Colonists* (London, 1864).

Pickstone, John (ed.). *Medical Innovations in Historical Perspective* (New York, 1992).

Porter, Dorothy and Roy Porter. *Patient's Progress; Doctors and Doctoring in Eighteenth-Century England* (Stanford, CA, 1989).

Porter, Roy. *Health For Sale: Quackery in England 1650–1850* (Manchester, 1989).

Porter, Roy. 'Laymen, Doctors, and Medical Knowledge in the Eighteenth-Century: the Evidence of the *Gentleman's Magazine*', in Roy Porter (ed.), *Patients and Practitioners: Lay Perceptions in Pre-Industrial Society* (Cambridge, 1985), 283–314.

Porter, Roy. 'The Physical Examination', in William Bynum and Roy Porter (eds), *Medicine and the Five Senses* (Cambridge, 1993), 179–97.

Porter, Roy. 'The Rise of Medical Journalism in Britain to 1800', in William Bynum, Stephen Lock and Roy Porter (eds), *Medical Journals and Medical Knowledge: Historical Essays* (London, 1992), 6–28.

Porter, Roy and Dorothy Porter, *In Sickness and in Health: the English Experience, 1650–1850* (London, 1988).

Porter, Roy and G. S. Rousseau. *Gout: the Patrician Malady* (New Haven, 1998).

Porter, T. *The Rise of Statistical Thinking 1820–1900* (Princeton, 1986).

Proudfoot, William Jardine. *Biographical Memoir of James Dinwiddie . . . Embracing Some Account of his Travels in China and Residence in India* (Liverpool, 1868).

Pulleyblank, E. and W. Beasley (eds). *Historians of China and Japan* (Oxford, 1961).

Puzey, Chauncy. 'Acupuncture', in Christopher Heath (ed.), *A Dictionary of Practical Surgery* (London, 1886), 24–6.

'Quarterly Periscope or, Spirit of the Public Journals . . . Acupuncture (Mr Churchill)', *Medico-Chirurgical Review, and Journal of Medical Science (Quarterly)*, 4 (1824): 956–7.

Ray, Kabita. *History of Public Health in Colonial Bengal, 1921–1947* (Calcutta: 1998).

Reingold, Nathan and Marc Rothenberg (eds). *Scientific Colonialism: a Cross Cultural Comparison* (Washington, D.C., 1989).

Reiser, Stanley J. *Medicine and the Reign of Technology* (Cambridge, 1978).

Reiser, Stanley J. and Michael Anbar (eds). *Medicine at the Bedside: Strategies for Using Technology in Patient Care* (Cambridge, 1984).

Renton, John. 'Observations on Acupuncture', *Edinburgh Medical and Surgical Journal*, 34 (1830): 100–1.

Report of the Speeches Delivered at the Public Dinner Given at Pennycuik to John Renton, Esq. Surgeon (Edinburgh, 1835).

'Retrospective of Foreign Medical Science and Literature, for the year 1819: I. Succinct Analysis of Foreign Periodical Literature', *London Medical Repository*, 13 (1820): 33–87.

'Retrospective Review: "Wilhelmi Ten Rhyne, MD. Transisalano. Daventriensis, *Dissertatio de Arthritide: Mantissa Schematica: De Acupunctura: Et Orationes Tres: I De Chymiae et Botaniae Antiquitate et Dignitate, II. De Physionomia, II De Monstris.* Lond. Impensis R Chiswell, 1689. 8vo" ', *North American Medical and Surgical Journal*, 1 (1826): 198–204.

'Review. *Medical and Surgical Cases: Selected During a Practice of Thirty-Nine Years –* By Edward Sutleffe', *Lancet* (22 April 1826): 102–9.

Richardson, Janet. 'Complementary Therapies on the NHS: the Experience of a New Service', *Complementary Therapies in Medicine*, 3 (1995): 153–7.

Richardson, Janet. *Complementary Therapy in the NHS: a Service Evaluation of the First Year of An Outpatient Service in a Local District General Hospital* (Lewisham, 1995).

Richardson, Janet and Ann-Marie Brennan. 'Complementary Therapy in the NHS: Service Development in a Local District Hospital', *Complementary Therapies in Nursing and Midwifery*, 1 (1995): 89–92.

Richardson, Ruth. *Death, Dissection and the Destitute: a Political History of the Human Corpse* (London, 1987).

Ritvo, Harriet. *The Mermaid and the Platypus and Other Figments of the Classifying Imagination* (Cambridge, MA, 1997).

Roberts, M. and Roy Porter. *Literature and Medicine During the 18th Century* (London, 1993).

Rosen, George. 'Lorenz Heister on Acupuncture: an Eighteenth Century View', *Journal of the History of Medicine and the Allied Sciences*, 30 (1975): 386–8.

Rosenberg, Charles. *The Care of Strangers: the Rise of America's Hospital System* (New York, 1987).

Rosenberg, Charles. 'Social Class and Medical Care in 19th Century America: the Rise and Fall of the Dispensary', in Judith Leavitt and Ron Numbers (eds),

Sickness and Health in America: Readings in the History of Medicine and Public Health, 2nd edition (Madison, 1985), 273–86.

Rosenberg, Charles. 'The Therapeutic Revolution: Medicine, Meaning, and Social Change in 19th Century America', in Judith Leavitt and Ron Numbers (eds), *Sickness and Health in America: Readings in the History of Medicine and Public Health*, 2nd edition (Madison, 1985), 39–52.

Rosenberg, Dorothy. 'Acupuncture and US Medicine: a Socio-Historical Study of the Response to the Availability of Knowledge' (PhD dissertation, University of Pittsburgh, 1977).

Rosenberg, Dorothy. 'Wilhelm Ten Rhyne's De Acupunctura: an 1826 Translation', *Journal of the History of Medicine and the Allied Sciences*, 34 (1979): 81–4.

Rousseau, G. S. and Roy Porter (eds). *The Ferment of Knowledge: Studies in Historiography of 18th Century Science* (Cambridge, 1980).

Rudwick, M. *The Great Devonian Controversy: the Shaping of Scientific Knowledge Among Gentlemanly Specialists* (Chicago, 1985).

Ruskin, John. *Works*, Vol. 28 (London, 1903–8).

Said, Edward. *Orientalism: Western Concepts of the Orient* (London, 1978).

Said, Edward. *Culture and Imperialism* (London, 1993).

Saks, Mike. 'The Flight from Science? The Reporting of Acupuncture in Mainstream British Medical Journals from 1800–1990', *Complementary Medical Research*, 5 (1991): 178–82.

Saks, Mike, *Professions and the Public Interest: Medical Power, Altruism and Alternative Medicine* (London, 1995).

Savitt, Todd. 'Black Health on the Plantation: Masters, Slaves, and Physicians,' in Judith Leavitt and Ron Numbers (eds), *Sickness and Health in America: Readings in the History of Medicine and Public Health*, 2nd edition (Madison, 1985), 313–30.

Schiebinger, Londa. *The Mind Has No Sex? Women in the Origins of Modern Science* (Cambridge, MA, 1989).

Scotus. 'Sciatica treated by Acupuncture, with Dr. Alison's Opinion on the Mode of its Operation', *Lancet* (19 May 1827): 190–1.

Shahine, Y. A. *The Arab Contribution to Medicine* (London, 1976).

Shapin, Steven and Simon Schaffer. *Leviathan and the Air-Pump: Hobbes, Boyle, and the Experimental Life: Including a Translation of Thomas Hobbes, Dialogus physicus de Natura Aeris, by Simon Schaffer* (Princeton, 1985).

Soulié de Morant, G. *L'Acupuncture Chinoise*, Vols 1 and 2 (Paris, 1939–41).

Spence, J. *Memory Palace of Matteo Ricci* (New York, 1984).

Spence, J. *The Search for Modern China* (New York, 1990).

Staunton, Sir George. *An Authentic Account of an Embassy from the King of Great Britain to the Emperor of China ... Taken Chiefly from the Papers of His Excellency the Earl of Macartney ... and of other Gentlemen in the Several Departments of the Embassy* (London, 1797).

Stephen, Leslie and Sidney Lee (eds), *The Dictionary of National Biography, from Earliest Times to 1900*, Vol. 7 (Oxford, 1949–50).

Sutleffe, Edward. *Medical and Surgical Cases; Selected During a Practice of Thirty-Eight Years*, Vol. 1 (London, 1824–5).

Teale, T. Pridgin. 'Clinical Essays, No. III. On the Relief of Pain and Muscular Disability by Acupuncture', *Lancet* (29 April 1871): 567–8.

Temple, Sir William. *Miscellanea. By a Person of Honour* (London, 1680).

Ten Rhyne, Wilhelm. *Dissertatio de Arthritide: Mantissa Schematica: De Acupunctura: Et Orationes Tres: I De Chymiae et Botaniae Antiquitate et Dignitate, II. De Physionomia, II De Monstris* (London, 1683).

Turner, Bryan. *Orientalism, Postmodernism and Globalism* (London, 1994).

Unschuld, Paul. *Medicine in China: a History of Ideas* (Berkeley, 1985).

Unschuld, Paul. *Medicine in China: a History of Pharmaceutics* (Berkeley, 1986).

Vicq D'Azyr, Felix (ed.). *Médecine. Contentant, l'Hygiêne; la Pathologie, la Séméiotique & la Nosologie; la Thérapeutique ou Matière Médicale; la Médecine Militaire; la Médecine Vétérinaire; la Médecine Légale; la Jurisprudence de La Médecine & de la Pharmacie; la Biographie Médicale. C'est-a-dire, les Vies des Médecins Célèbres, avec des Notices de leurs Ouvrages,* in Diderot & d'Alembert, (eds), *Encyclopédie Méthodique* (Paris, 1792).

Vogel, Morris and Charles Rosenberg (eds). *The Therapeutic Revolution* (New York, 1979).

Waddington, I. 'General Practitioners and Consultants in Early Nineteenth-Century England: the Sociology of an Intra-Professional Conflict', in J. H. Woodward and D. Richards (eds), *Health Care and Popular Medicine in Nineteenth Century England* (New York, 1977).

Walkowitz, Judith R. *Prostitution and Victorian Society: Women, Class, and the State* (Cambridge, 1980).

Wallis, R. (ed.). *On the Margins of Science: the Social Construction of Rejected Knowledge* (Keele, 1979).

Wansbrough, T. W. 'Case of Rheumatism Successfully Treated by Acupuncturation', *Lancet* (21 June 1828): 366–7.

Wansbrough, T. W. 'Acupuncturation', *Lancet* (30 September 1826): 846.

Warner, John Harley. 'The Idea of Southern Medical Distinctiveness: Medical Knowledge and Practice in the Old South', in Judith Leavitt and Ron Numbers (eds), *Sickness and Health in America: Readings in the History of Medicine and Public Health*, 2nd edition (Madison, 1985), 53–70.

Webster, Charles. *The Great Instauration: Science, Medicine and Reform, 1626–1660* (London, 1975).

Willis, John Jr. 'European Consumption and Asian Production in the Seventeenth and Eighteenth Centuries', in John Brewer and Roy Porter (eds), *Consumption and the World of Goods* (London, 1993), 133–47.

Winter, Alison. *Mesmerized: Powers of Mind in Victorian Britain* (Chicago, 1998).

Wilson, John. *Medical Notes on China* (London, 1846).

Wilson, Ming and John Cayley (eds). *Europe Studies China* (London, 1995).

Wong Chimin and Wu Lien-te. *History of Chinese Medicine: Being a Chronicle of Medical Happenings in China from Ancient Times to the Present Period* (Tientsin, 1932).

Woodward, J. H. and D. Richards (eds). *Health Care and Popular Medicine in Nineteenth Century England* (New York, 1977).

Wotton, William. *Reflections upon Ancient and Modern Learning* (London, 1694).

Yolton, John W. et al. (eds). *The Blackwell Companion to the Enlightenment* (Oxford, 1991).

Young, R. *Darwin's Metaphor: Nature's Place in Victorian Culture* (Cambridge, 1985).

Index

References to illustrations are printed in **bold** type. I have included one general and several specific entries for the term 'acupuncture'.